DNA Damage by Auger Emitters

DNA Damage by Auger Emitters

Edited by
K.F. Baverstock
MRC Radiobiology Unit, Harwell, UK

and **D.E. Charlton**
Concordia University, Montreal, Canada

Taylor & Francis
London · New York · Philadelphia
1988

UK	Taylor & Francis Ltd, 4 John St, London WC1N 2ET
USA	Taylor & Francis Inc., 242 Cherry St, Philadelphia, PA 19106-1906

Copyright © Taylor & Francis 1988

All rights reserved. No part of this publication may be reproduced, stored in a retrieval system, or transmitted, in any form or by any means, electronic, electrostatic, magnetic tape, mechanical, photocopying, recording or otherwise, without the prior permission of the copyright owner.

British Library Cataloguing in Publication Data

DNA damage by auger emitters.
 1. Organisms. DNA. Damage
 I. Baverstock, K.F. II. Charlton, D.E.
 574.87'3282

ISBN 0-85066-406-3

Library of Congress Cataloging in Publication Data is available

Cover artwork courtesy of Dr. Synnove Sundell-Bergman

Printed in Great Britian by Taylor & Francis (Printers) Ltd, Basingstoke

CONTENTS

Abbreviations	viii
Preface	ix
Acknowledgements	xi
Participants list	xiii

POSITIONAL EFFECTS OF AUGER DECAYS IN MAMMALIAN CELLS IN CULTURE
A.I. Kassis, R.W. Howell, K.S.R. Sastry & S.J. Adelstein
.... 1

INTERNAL AUGER-EMITTERS: EFFECTS ON SPERMATOGENESIS AND OOGENESIS IN MICE
D.V. Rao, V.B. Mylavarapu, K.S.R. Sastry & R.S. Howell
.... 15

DOSIMETRY OF AUGER-EMITTERS: PHYSICAL AND PHENOMENOLOGICAL APPROACHES
K.S.R. Sastry, R.W. Howell, D.V. Rao, V.B. Mylavarapu, A.I. Kassis, S.J. Adelstein, H.A. Wright, R.N. Hamm & J.E. Turner
.... 27

HIGH-LET ENDORADIOTHERAPY: *IN VIVO* STUDIES WITH A METABOLICALLY-DIRECTED AUGER EMITTING ANTICANCER DRUG IN A MURINE TUMOUR MODEL
I. Brown, R.N. Carpenter & J.S. Mitchell
.... 39

RADIOCHEMOTHERAPY WITH ^{125}I-5-IODO-2-DEOXYURIDINE
K.D. Bagshawe
.... 51

DNA DAMAGE BY AUGER EMITTERS 55
R.F. Martin, B.J. Allen, G. d'Cunha,
R. Gibbs, V. Murray & M. Pardee

CELL KILLING AND DNA DOUBLE-STRAND BREAKAGE 69
BY DNA-ASSOCIATED ^{125}I-DECAY OR X-IRRADIATION;
IMPLICATIONS FOR RADIATION ACTION MODELS
I.R. Radford & G.S. Hodgson

NEW CONCEPTS OF DNA FUNCTIONING AS REVEALED 81
BY THE LETHAL EFFECT OF ^{64}Cu AND ^{67}Cu DECAYS
S. Apelgot & E. Guille

CALCULATION OF SINGLE AND DOUBLE STRAND 89
DNA BREAKAGE FROM INCORPORATED ^{125}I
D.E. Charlton

THE POSSIBLE ROLE OF SOLITONS IN ENERGY 101
TRANSFER IN DNA: THE RELEVANCE OF STUDIES
WITH AUGER EMITTERS
K.F. Baverstock & R.B. Cundall

DOUBLE STRAND BREAKAGE IN DNA PRODUCED 111
BY THE PHOTOELECTRIC INTERACTION WITH
INCORPORATED 'COLD' BROMINE
J.L. Humm & D.E. Charlton

AN ADDITIONAL ENHANCEMENT IN BrdU-LABELLED 123
CULTURED MAMMALIAN CELLS WITH MONOENERGETIC
SYNCHROTRON RADIATION AT 0.09nm: AUGER EFFECT
IN MAMMALIAN CELLS
H. Ohara, K. Shinohara, K. Kobayashi & T. Ito

EFFECTS OF AUGER CASCADES OF BROMINE INDUCED 135
BY K-SHELL PHOTOIONIZATION ON PLASMID DNA,
BACTERIOPHAGES, *E.COLI* AND YEAST CELLS
H. Maezawa, K. Hieda, K. Kobayashi & T. Ito

RADIOTOXIC EFFECTS OF ^{125}I-LABELLED THYROID 147
HORMONES WITH AFFINITY TO CELLULAR CHROMATIN
S. Sundell-Bergman, K.J. Johanson &
G. Ludwikow

GENE AMPLIFICATION IN CHINESE HAMSTER EMBRYO 159
DELLS BY THE DECAY OF INCORPORATED IODINE-125
C. Lücke-Huhle, A. Ehrfeld & W. Rau

LETHAL MUTAGENIC AND DNA-BREAKING EFFECTS OF 169
DECAYS OF IODINE-125 UNIFILARLY INCORPORATED
INTO THE DNA OF MAMMALIAN CELLS
Y. Fujiwara & N. Miyazaki

CYTOTOXICITY OF ^{125}I DECAY PRODUCED LESIONS 181
IN CHROMATIN
L.S. Yasui, A.S. Paschoa, R.L. Warters
& K.G. Hofer

Index   191

Abbreviations

A number of abbreviations are used in the texts of these papers and as far as possible a self consistent system has been used throughout. Most are self explanatory but it may be helpful to outline the system employed in designating the DNA bases and their derivatives. The abbreviations are derived as follows:-

<u>Bases</u> designated by upper case initial letters; e.g., T for thymine.

<u>Deoxy</u> derivatives: base preceded by a lower case "d"; e.g., dT for thymidine.

<u>Halide</u> derivatives; base preceded by the usual halide abbreviation; e.g., BrdC for bromo-deoxy cytidine.

<u>Phosphate</u> derivatives: base followed by upper case letters to indicate mono "M", di "D", and tri "T", and then "P"; e.g., BrdUTP for bromo-deoxy uridine triphosphate.

PREFACE

In 1975 an International conference on the biological consequences of internally incorporated radionuclides was held in Julich, Germany, and included 10 papers on the effects of ^{125}I and 3 papers on the possible application to cancer therapy of Auger emitting isotopes[1]. To the best of our knowledge there has not, until now, been an International meeting devoted solely to the problems and usage associated with Auger emitters.

In 1985 the Medical Research Council[2] of Great Britain addressed the question of how to assess the dosimetry of radiopharmaceuticals (many of which are Auger emitters) in such a way that the risks of such exposure is truely reflected. The problem identified here is the short range of much of the Auger emission and hence the question, over how large a volume is it appropriate to calculate the absorbed dose?

Thus two important medical applications for Auger emitters emerge - one, a 'promise' for the future of an effective means of controlling cancer, the other an existing 'problem' in assessing risk in nuclear medicine. At the root of both is that characteristic of Auger decay, the highly selective irradiation of the volume immediately surrounding the emitter, which on the one hand poses the problem in nuclear medicine and on the other holds out the promise of selective attack on cancer cells, what was called by Professor Feinendegen 'molecular surgery'.

---

[1] Current topics in Radiation Research Quarterly (1977) $\underline{12}$.
[2] International Journal of Radiation Biology (1986) $\underline{50}$, 555-567.

Interestingly, Auger emitters, as is amply demonstrated in these pages, are an important tool in probing basic biological mechanisms and may therefore ultimately provide the solutions to the questions they themselves raise in medicine. Seen in this light the diversity of discipline represented in this collection of papers has a unifying basis.

The papers published here are ordered as they were presented at the Workshop and start with the medical aspects, both diagnostic and therapeutic. Understanding of the nature of the dosimetric problem and of how effective cancer therapy is to be developed requires an understanding of the effects of Auger emitters at the molecular level and the following five papers are broadly devoted to this aspect.

An important new tool in Auger science is the Synchrotron, the emissions from which can stimulate Auger cascades in 'cold' atoms. Work in this field from Japan is described along with theoretical studies which assist in the interpretation of experimental data. Finally, important studies at the cellular level with incorporated Auger emitters, including the exploitation of molecular biological techniques, provide one of the major growth points of the subject.

We would hope that those not presently using the Auger emitter tool will draw inspiration from the work reported here and see applications in their own fields of study thus further broadening the diversity already inherent in this subject.

K.F. Baverstock
D.E. Charlton
Editors

ACKNOWLEDGEMENTS

The editors wish to express their gratitude to the following who helped to ensure the success of the meeting and preparation of the proceedings.

Referees
Duncan Ackery, Alan Edwards, Dudley Goodhead, Alice Harrison, John Humm, Hooshang Nikjoo, John Thacker.

Preparation of typescript
Louise Ashby, Caroline Beaney, Vikki Collyer, Ann Lewis.

Proof readers
Cindy McIntyre, Angela Wilson.

Session Chairmen
Duncan Ackery, Jim Adelstein, Takashi Ito.

Speakers at the Preview and Summary sessions
Jim Adelstein, Takashi Ito, Roger Martin and Synnove Sundell-Bergman.

Gratitude is also extended to Amersham International p.l.c. for a donation and to Adrienne Jackson and Ann Pack of Norsk Data for assistance with the preparation of the typescript.

The participants at the workshop on DNA Damage by Auger Emitters, at Charney Manor, Oxfordshire, on 17 July 1987.

PARTICIPANTS

Ackery, D., Dept. of Nuclear Medicine, Southampton University, Southampton General Hospital, Southampton. SO9 4XY, U.K.

Adelstein, S.J., Harvard Medical School, 25 Shattuck Street, Boston, Massachussetts MA 02115 U.S.A.

Al-Kazwini, A.T., MRC Radiobiology Unit, Chilton, Didcot, Oxon. OX11 ORD, U.K.

Apelgot, S., Institut Curie, Section de Physique et Chemie, 11, rue Pierre-et-Marie Curie, 75321 Paris Cedex 05, France.

Bagshawe, K.D., Dept. of Medical Oncology, Charing Cross Hospital, LONDON. W6 8RF, U.K.

Baverstock, K.F., MRC Radiobiology Unit, Chilton, Didcot, Oxon. OX11 ORD, U.K.

Brown, I., The Research Laboratories, Radiotherapeutic Centre, Cambridge University School of Clinical Medicine, Addenbrooke's Hospital, Cambridge CB2 2QQ, U.K.

Carpenter, R.N., The Research Laboratories, The Radiotherapeutic Centre, Addenbrookes Hospital, Hills Road, Cambridge CM2 2QQ, U.K.

Charlton, D.E., Department of Physics, Sir George William's Campus, 1455 de Maisonneuve, Boulevard West, Montreal, Quebec H3G 1M8, Canada.

Cundall, R.B., MRC Radiobiology Unit, Chilton, Didcot, Oxon. OX11 ORD, U.K.

Denison, L., Peter MacCullum Cancer Institute, 481 Little Lonsdale Street, Melbourne 3000, Australia.

Edwards, A.A., National Radiological Protection Board, Chilton, Didcot, Oxon. OX11 ORD, U.K.

Ellis, R., 2 Bladon Close, Oxford. OX2 8AD, U.K.

Fairchild, R.G., Medical Dept., Brookhaven National Laboratories, Upton, N.Y. 11943, U.S.A.

Fujiwara, Y., Kobe University School of Medicine, Dept. of Radiation Biophysics, Kusunoki-cho 7-5-1, Chuo-Ku, Kobe 650, Japan.

Goodhead, D.T., MRC Radiobiology Unit, Chilton, Didcot, Oxon, OX11 ORD, U.K.

Halpern, A., KFA Juelich, Institut für Chemie 1 (Nuklearchemie), D-5170 Juelich, W. Germany.

Hamm, R.N., Oak Ridge National Laboratory, Oak Ridge TN 37831-6123, U.S.A.

Harrison, A., MRC Radiobiology Unit, Chilton, Didcot, Oxon, OX11 ORD, U.K.

Hieda, K., Biophysics Laboratory, Faculty of Science, Rikkyo University, Toshimaku, Tokyo 171, Japan.

Howell, R.W., Department of Physics and Astronomy, University of Massachusetts, Amherst MA 01003, U.S.A.

Humm, J.L., MRC Radiobiology Unit, Chilton, Didcot, Oxon, OX11 ORD, U.K.

Ito, T., Institute of Physics, University of Tokyo, Komaba, Meguroku, Tokyo 153, Japan.

Johanson, K.J., The Swedish University of Agricultural Sciences, Dept. of Radioecology, S - 750 07 Uppsala, Sweden.

Kaneko, I., RIKEN, The Institute of Physical and Chemical Research, Wako, Saitama 351-01, Japan.

Kassis, A.I., Harvard Medical School, SWRL, 50 Binney Street, Boston, MA 02115, U.S.A.

Kobayashi, K., Photon Factory, National Laboratory for High Energy Physics, Oho, Ibaraki 305, Japan.

Lücke-Huhle, C., Kernforschungszentrum Karlsruhe GmbH, Institut für Genetik und für Toxikologie von Spaltstoffen, Postfach 3640, D-7500 Karlsruhe 1, W. Germany.

Ludvikoff, G., Dept. of Radioecology, Swedish University Agric. Sci., Box 7031 750 07 Uppsala, Sweden.

Maezawa, H., Dept. Radiation Oncology, Tokai U.V., School of Medicine, Bohseidai, Isehara, Kanagawa 259-11, Japan.

Martin, R., Molecular Science Group, Peter MacCallum Cancer Institute, 481 Little Lonsdale Street, Melbourne, Victoria 3000, Australia.

McIntyre, C., MRC Radiobiology Unit, Chilton, Didcot, Oxon. OX11 ORD, U.K.

Mylavarapu, V.B., University of Medicine and Dentistry of New Jersey, New Jersey Medical School, 100 Bergen Street, Newark, New Jersey 07103, U.S.A.

Nikjoo, H., MRC Radiobiology Unit, Chilton, Didcot, Oxon. OX11 ORD, U.K.

Ohara, H., Institute of Physics, University of Tokyo, Komaba, Meguroku, Tokyo 153 JAPAN.

Radford, I.R., Molecular Sciences Group, Peter MacCallum Cancer Institute, 481 Little Lonsdale Street, Melbourne, Victoria 3000, Australia.

Rao, D.V., University of Medicine and Dentistry of New Jersey, University Hospital, 100 Bergen Street, Newark, New Jersey 07103, U.S.A.

Raù, W., Kernforschungszentrum Karlsruhe GMBH, Institut für Genetik und für Toxikologie von Spaltstoffen, Postfach 3640, D-7500 Karlsruhe 1, W. Germany.

Sastry, K.S.R., Department of Physics and Astronomy, University of Massachusetts, Amherst, MA 01003 U.S.A.

Sundell-Bergman, S., Department of Radioecology, Swedish University, Agricultural, Science, Box 7031, 750 07 Uppsala, Sweden.

Wilson, A., MRC Radiobiology Unit, Chilton, Didcot, Oxon. OX11 ORD, U.K.

Yasui, L.S., Radiology Dept., University of Utah Medical Centre, Salt Lake City, Utah 84132, U.S.A.

Younis, A.R.S., St. Andrews University, Physics Department, St. Andrews, Fife, KY16 9SS, U.K.

POSITIONAL EFFECTS OF AUGER DECAYS IN MAMMALIAN CELLS IN CULTURE

A.I. Kassis,[1] R.W. Howell,[2] K.S.R. Sastry,[2] & S.J. Adelstein[1]

[1]Department of Radiology
Harvard Medical School
50 Binney Street, Boston, Massachusetts, USA

[2]Department of Physics and Astronomy
University of Massachusetts
Amherst, Massachusetts, USA

ABSTRACT

Over the past several years, we have been interested in determining the radiotoxicity of subcellularly localized Auger electron emitters. We have examined the toxicity of DNA-incorporated ^{125}I and ^{77}Br, DNA-intercalated ^{125}I, DNA-bound ^{195m}Pt, cytoplasmic ^{125}I and ^{75}Se, extracellular ^{125}I$^-$ and ^{77}Br$^-$, as well as ^{201}Tl and ^{51}Cr located within several compartments of the cell. Intracellular and extracellular concentration, exposure time, and cell cycle effects have been studied. Except in the case of iodide and bromide, the radiolabelled materials are concentrated by cells, and their radiotoxicity depends on the subcellular distribution of the Auger emitter. Our results emphasize the importance of the *in vitro* biological approach, both in understanding the effects of the Auger emitters and in estimating the dose delivered to the radiosensitive sites in the cell.

INTRODUCTION

With respect to both their diagnostic and therapeutic roles, it is important to understand the dosimetry of radioisotopes. Many radionuclides previously or currently used in nuclear medicine, such as selenium-75, thallium-201, iodine-123, chromium-51, indium-111, and technetium-99m, decay by electron capture (EC) and/or internal conversion (IC). Both EC and IC cause vacancies in inner electron orbits that are immediately filled by electrons from higher shells with the emission of x-rays or the ejection of electrons. The electron transition process may be repeated if the electron which has moved to fill the inner shell vacancy did not come from the outermost shell, and thus a complex series of vacancy cascades resulting in the release of numerous Auger and Coster-Kronig (CK) electrons may arise from the original event.

Since the energy of Auger and CK electrons is quite low (in the range of 20 eV to 250 keV), the distances these electrons traverse in biological matter are very short (1 to 10 nm). The values of linear energy transfer (LET) for these electrons range from 10 to 25 keV/μm, and thus the energy is deposited in the immediate vicinity of the site of the initial event of inner shell ionization. However, the radiobiological effects following the decay of such radionuclides are not readily predictable because of their dependence on (a) the localization of the radionuclide relative to the radiosensitive target(s) within nuclear DNA; (b) the energy, number, interval of release, etc., of the Auger electrons; and (c) the patterns of energy deposition of these electrons at the microscopic level. Furthermore, classical dosimetry of tissue-incorporated radionuclides, based on the simplifying assumptions inherent in the MIRD Schema (Loevinger & Berman, 1976) and the ICRU procedures (ICRU Report 32, 1979), is expected to lead to gross underestimation of the radiation doses to cells in the case of intracellularly localized Auger emitters.

In view of these considerations, we have been investigating the kinetics of uptake and retention, the intracellular distribution, and the radiotoxicity of several Auger electron emitters of biomedical interest using in vitro model systems. We have compared the biological data with the calculated localized energy deposition by the Auger electrons using theoretical estimates of the electron spectra for the radionuclides. In this paper we present the results obtained with a single cultured mammalian cell line. Theoretical dose calculations will be discussed in the following communication (Sastry et al., 1987).

MODEL CELL SYSTEM

For these studies we have employed an in vitro model cell system that can be easily manipulated under strictly controlled conditions. The cell line (Chinese hamster V79 lung fibroblasts) has the following characteristics: (a) it is stable from generation to generation; (b) it grows readily in monolayer or suspension; (c) it has a high plating efficiency; (d) it forms colonies, thereby providing an easy end point for scoring loss of reproductive capacity.

The V79 cells used in most of our studies, when suspended in complete medium, have a diameter of 10.3 μm and a nuclear diameter of ~8.0 μm (Kassis et al., 1980). Their diploid chromosome number is 22. Heussen et al. (1987) have shown that the molecular weight of the chromatin is 10.6×10^6 with a size range 6.6 to 21.7×10^6 Daltons. The repeat length of the chromatin has been found to be 194 ± 3 base pairs.

RADIONUCLIDE SELECTION

From an experimental point of view, a radionuclide whose intracellular concentration within various subcompartments of a cell can be determined is ideal. In our early work we used readily available Auger-electron-labelled molecules and examined their intracellular distribution and cytotoxicity. More recently, we have been custom-synthesizing labelled molecules to target the Auger electron emitter to specific compartment(s) within the cell.

METHODS

Measurement of radioactive cell content

Logarithmically growing monolayers of cells are suspended in various concentrations of the radionuclide and incubated for up to 18 hours (Kassis et al., 1983). Both the trichloroacetic-acid (TCA)-precipitable cell-incorporated (DNA, RNA, and protein) activity and the extracellular radioactive concentrations are determined. When necessary, cells are synchronized by mitotic shakeoff.

Due to the diffusible nature of certain radiopharmaceuticals, we have developed a method for the rapid separation of cells from radioactive media (Kassis & Adelstein, 1980). In essence, an aliquot of cells in aqueous medium is layered onto fetal bovine serum in microfuge tubes; the tubes are spun for 1 minute, frozen in ethanol-dry ice, their tips sliced off, and the radioactivity assayed in a scintillation counter. The technique is simple, rapid, and highly reproducible. It allows us to work with a small volume of cells in suspension. Only small amounts of radioactivity, therefore, are required, limiting both the expenditure for radionuclides and the exposure of personnel.

Determination of intracellular radionuclide distribution

Following incubation with the radionuclide, V79 cells are washed, lysed under hypotonic conditions, and the nuclei isolated (Hymer & Kuff, 1964). Mitochondria are sedimented by centrifugation of the cytoplasmic fraction. The TCA- and/or guanidine HCl (DNA)-precipitable activity is determined.

Figure 1. Clonogenic survival of mammalian cells following DNA incorporation of ^{125}I, ^{131}I, and ^{3}H.

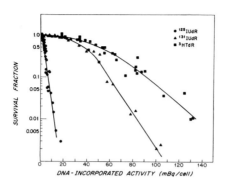

Cell survival assay

Cell survival is assayed by incubating a sufficient number of cells in plastic flasks to yield 30 to 250 colonies seven days after exposure to the radionuclide. The colonies are fixed with Bouin's fixative, stained with trypan blue, counted, and the mean survival fraction (S/So) determined (Kassis et al., 1980). The ability of a single cell to form colonies of $\geq$ 50 cells indicates survival.

It is essential when determining the toxicity of a radionuclide to investigate the chemotoxicity of the labelled compound at the same molar concentrations. In all the studies described here, we have examined cell survival in the presence of either the compound after the radionuclide has decayed or the nonradioactive parent compound when the physical half-life of the radionuclide is too long to permit examination of the decayed product(s).

Dosimetric calculations

The cumulative dose to the cell nucleus is determined as described by Kassis et al., (1980, 1983, 1985b, 1987a).

RADIOTOXICITY OF AUGER ELECTRON EMITTERS

Our initial work examined the basis for the exquisite radiotoxicity observed with iodine-125 incorporated into DNA as [^{125}I]iododeoxyuridine (^{125}IdU) (Bradley et al.,

Figure 2. Clonogenic survival of V79 cells following 18-hr incubation in (A) [^{75}Se]selenomethionine and (B) [^{77}Br]-bromodeoxyuridine.

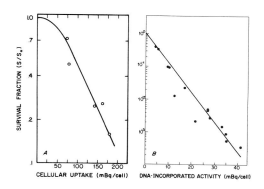

1975; Chan et al., 1976). The survival curve elicited by this compound is of the high LET type with no shoulder (Figure 1). In these studies, we also compared the radiotoxicity of this radionuclide with that of two beta emitters, ^{3}H (tritiated thymidine) and ^{131}I ([^{131}I]iododeoxyuridine). Following the DNA incorporation of these beta emitters, the survival curves obtained are of the low LET type with a definite shoulder.

We extended these studies to two other Auger electron emitters, ^{77}Br and ^{75}Se. As bromodeoxyuridine (BrdU), ^{77}Br is incorporated into DNA (similarly to ^{125}I in iododeoxyuridine and ^{3}H in thymidine); as selenomethionine, ^{75}Se is incorporated into cytoplasmic proteins. We assumed that the rate of uptake with time would be linear and that once incorporated, the radionuclides would remain intracellular, thereby permitting determination of their biological half-lives. The results (Figure 2) fulfilled our expectations and indicated that the decay of an Auger emitter (^{75}Se) within the cytoplasm generates a survival curve of the low LET type with a D_{37} of 145 mBq/cell (Kassis et al., 1980), while DNA incorporation (^{77}Br) results in a toxicity curve of the high LET type and a D_{37} of 4.8 mBq/cell (Kassis et al., 1982).

Intracellular concentration effects

In nuclear medicine, chromium-51 is frequently employed

as a cellular label to measure lifetime and kinetics and thallium-201 to measure myocardial perfusion in patients with coronary artery disease. Both radionuclides decay by Auger electron emissions. We were concerned that classical dosimetry would underestimate their toxicity because of the non-uniform distribution of low energy electrons in tissue and because of potential quality factor effects if the radionuclides were concentrated in the cell nucleus.

The rate of uptake and retention of ^{201}Tl (as thallous) (Adelstein, 1980). The uptake is brisk and the discharge extremely rapid as would be expected for a monovalent cation not bound by specific cell components. The cell survival curve is of the low LET type with a D_{37} of 45 mBq/cell (Kassis et al., 1983). When the active uptake of this radionuclide is specifically inhibited by incubating the cells in the presence of ouabain (a drug that interferes with the energy-dependent sodium-potassium ATPase pump), the survival fraction of ^{201}Tl-incubated cells increases 400-fold at higher ouabain concentrations

Figure 3. A. Uptake and clonogenic survival of V79 cells following 18-hr incubation with ^{201}TlCl in presence of various concentrations of ouabain. B. Intracellular distribution of ^{51}Cr in V79 cells following 18-hr incubation in single radioactive concentration of sodium [^{51}Cr]chromate.

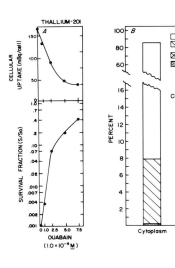

Figure 4. Growth of V79 cells as function of time following 18-hr incubation in (A) ^{51}Cr and (B) ^{125}IdU.

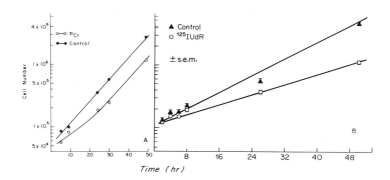

(Figure 3A), demonstrating the importance of intracellular ^{201}Tl localization for the manifestation of its cytocidal effects. The low-LET-type survival curve suggests that ^{201}Tl is distributed throughout the cell and that the toxicity observed depends principally on concentration effects.

For sodium [^{51}Cr]chromate, whose intracellular distribution was unknown in V79 cells, little leakage is observed when the cells are removed from the radioactive medium. Cell fractionation techniques indicate that although ~14% of the cellular activity is associated with the nucleus, only ~2% is bound to the DNA (Figure 3B); consequently, it is not surprising to obtain a cell survival curve of the low LET type (Kassis et al., 1985b) with a D_{37} of 230 mBq/cell.

Division delay and cell cycle effects

As experimental data on the radiotoxicity of Auger emitters have accumulated, we have realized the importance of other biological parameters on the observed cytotoxicity. For instance, it has become apparent that when calculating the cumulative dose to cells after exposure to or incorporation of radionuclide, since it is the actual intracellular content that is important, cell doubling time and the progression through the cell cycle must be taken into account. Moreover, the relative radiosensitivities in various phases of the cell cycle must also be examined.

Experimental observations made during our studies on the radiotoxicity of chromium-51 in V79 cells suggested that the cell doubling time (normally 9 hours) increases following intracellular incorporation of the radionuclide. Cells pre-exposed to 370 kBq/ml (≡ 190 mBq/cell) of ^{51}Cr show a biphasic growth curve: an early phase with a doubling time of 14 hours lasting for 28 hours, and a later phase whose slope parallels that of the control (Figure 4A), i.e., returns to a doubling time of 9 hours. This phenomenon is also observed following ^{125}IdU incorporation into DNA (Figure 4B). In this instance, the doubling time of V79 cells pre-exposed to only 3.7 kBq/ml (≡ 20 mBq/cell) increases to ~15 hours, indicating that the incorporation of this Auger emitter into the DNA of dividing mammalian cells causes a marked decrease in cell growth rate (Kassis et al., 1987a). Similar conclusions have been reached by Schneiderman & Hofer, (1980) and Warters & Hofer, (1977).

To examine cell cycle effects, we have compared the survival of synchronized Chinese hamster ovary (CHO) cells after x-irradiation in early and mid S phase of the cell cycle with that obtained in cells pulsed with ^{125}IdU during the same periods (Kassis et al., 1986). These

Figure 5. Clonogenic survival of V79 cells exposed to various radioactive concentrations of ^{125}I-DR for 18 hr.

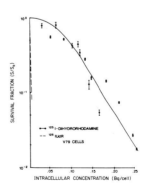

preliminary results indicate that CHO cells radiolabelled in mid S phase are more susceptible to radiation-induced sterilization by ^{125}I decay than cells radiolabelled in early S phase. No differential radiosensitivity between the two phases is found with x-irradiation. These findings are in agreement with a recent study (Yasui et al., 1985) which indicates that nuclear DNA is not homogeneous as a target with respect to radiation-induced cell death.

Toxicity in relation to cellular localization

We have begun to synthesize compounds radiolabelled with Auger electron emitters that localize in various subcellular compartments. Iododihydrorhodamine (I-DR), a derivative of rhodamine 123 which is a fluorescent permeant cationic laser dye that selectively stains mitochondria, has been recently synthesized and characterized in our laboratory (Kinsey et al., 1987). When radiolabelled with iodine-125, ~96% of ^{125}I-DR localizes in the cytoplasm of V79 cells, and its radiotoxicity has the characteristics expected of an Auger emitter restricted to this region of the cell (Figure 5), i.e., producing a survival curve of the low LET type with a D_{37} of 109 mBq/cell at an extracellular concentration of 1.07 MBq/ml. This is in contrast to the high LET type of survival curve obtained following the DNA incorporation of ^{125}IdU with a D_{37} of

Figure 6. Clonogenic survival of V79 cells incubated for 18 hr in varying nonradioactive concentrations of IdU (A) or TdR (B) in absence (o) or presence (●) of 9.25 kBq/ml ^{125}IdU.

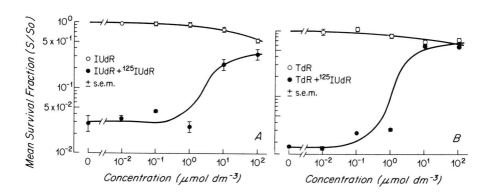

1.5 mBq/cell (Kassis et al., 1987a).

In our earlier work with ^{125}IdU, cells were incubated with different radioactive concentrations of the radioactive nucleoside to vary the amount of iodine-125 incorporated into DNA. Since variable concentrations of extracellular iodine-125 could have contributed to the observed radiotoxicity, we have varied the TCA-precipitable iodine-125 content of DNA while maintaining a constant extracellular concentration of ^{125}IdU by competitive inhibition with nonradioactive IdU or thymidine (Kassis et al., 1987a). Our results, demonstrating that the toxicity depends on the amount of iodine-125 incorporated into the nucleus, argue against any significant contribution from extranuclear iodine-125 located in the medium (Figure 6). The work of Yasui and Hofer, (1986) in which the mitochondrial, but not nuclear, DNA incorporation of IdU was inhibited without altering the cytocidal effects of the radionuclide also supports these conclusions.

None of these studies answer the following two questions: (a) Is the incorporation of an Auger electron emitter into the DNA of mammalian cells a prerequisite for the production of high-LET-type dose survival curves? (b) Does the RBE change when the Auger emitter is bound to or intercalated with DNA? In our attempts to resolve these issues, we have recently synthesized 3-acetamido-5-[^{125}I]iodoproflavine (^{125}IAP), a fluorescent dye that intercalates with and selectively stains nuclear DNA, and trans-^{195m}Pt, a radiopharmaceutical that binds to nuclear DNA. Cells exposed to these compounds do, in fact, fact, exhibit survival curves of the high LET type with D_{37}'s of 3.50 and 1.2 mBq/cell, respectively (Figure 7).

The cumulated dose (D_N) to the nucleus at 37% survival from ^{125}I following DNA incorporation is 0.80 Gy (Kassis et al., 1987a). When compared with the corresponding value of 5.80 Gy from 250 kVp x-rays for the same cell line (Kassis et al., 1980), an RBE (relative biological effectiveness) of 7.3 is obtained for ^{125}IdU (Table 1). For ^{77}BrdU, D_N is 0.89 Gy with a corresponding RBE value of 6.5 (Kassis et al., 1987a). On the other hand, the RBE values for ^{125}IAP, which intercalates with DNA, and for trans-^{195m}Pt, which forms adducts with DNA, is 4.8 for both radionuclides. In contrast, ^{125}I localized in the nucleus but assumed not to be bound to DNA (i.e., the 3.7% fraction of ^{125}I-DR) is much less efficient at cell killing and results in an RBE of ~1.3 (Kassis et al., 1987b). These results support the notion that the efficiency of an Auger electron emitter in causing cytocidal effects is strongly dependent on its proximity to nuclear DNA.

Figure 7. Clonogenic survival of V79 cells following 18 hr incubation in various radioactive concentrations of (A) ^{125}IAP and (B) trans-^{195m}Pt.

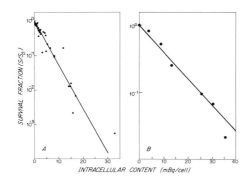

Table 1. Comparison of nuclear dose (D_N) and relative biological effectiveness (RBE) of several Auger electron emitters in V79 cells.

Radiochemical	D_N (Gy)	RBE
^{125}IdU	0.80	7.3
^{77}BrdU	0.89	6.5
Trans-^{195m}Pt	1.22	4.8
^{125}IAP	1.21	4.8
^{125}i-dr	4.46	1.3

CONCLUSION

We have examined several Auger electron emitters and attempted to determine whether their radiotoxicity in cultured mammalian cells can substantiate the overall theoretical expectations. In general, the experimental data have validated the physical interpretation. Through the use of a well characterized mammalian cell line and the synthesis of site-directed radiolabelled compounds we have demonstrated that: (a) the extreme radiotoxicity of iodine-125 incorporated into DNA as the halogenated pyrimidine nucleotide is independent of specific activity and is shared with another Auger-emitting radiohalogen, bromine-77; (b) when decaying in close proximity to DNA, the radiotoxic effects of iodine-125 and platinum-195m, while somewhat milder than DNA-incorporated ^{125}I and ^{77}Br, also produce a high-LET-type survival curve and have a high RBE; (c) iodine-125 incorporated into DNA in mid S phase appears to be more radiotoxic than iodine-125 incorporated into DNA in early S phase, and there is little difference in radiation sensitivity to x-rays between early and mid S phase cells in our system; (d) Auger-electron-emitting radionuclides that are concentrated by cells and not bound to or incorporated into DNA produce low-LET-type survival curves, whether localized in the cytoplasm or more diffusely distributed within the cell.

This work is supported by USPHS grants CA15523 and CA32877.

REFERENCES

Bradley, E.W., Chan, P.C. & Adelstein, S.J., 1975, Radiation Research, 64, 555.

Chan, P.C., Lisco, E., Lisco, H., & Adelstein, S.J., 1976, Radiation Research, 67, 332.

Heussen, C., Nackerdien, Z., Smit, B.J. & Bohm, L., 1987, Radiation Research, 110, 84.

Hymer, W.C. & Kuff, E.L., 1964, Journal of Histochemistry and Cytochemistry, 12, 359.

ICRU Report 32, 1979, Methods of Assessment of Absorbed Dose in Clinical Use of Radionuclides (International Commission on Radiation Units and Measurements).

Kassis, A.I. & Adelstein, S.J., 1980, Journal of Nuclear Medicine, 21, 88.

Kassis, A.I., Adelstein, S.J., Haydock, C. & Sastry, K.S.R., 1980, Radiation Research, 84, 407.

Kassis, A.I., Adelstein, S.J., Haydock, C. & Sastry, K.S.R., 1983, Journal of Nuclear Medicine, 24, 1164.

Kassis, A.I., Adelstein, S.J., Haydock, C., Sastry, K.S.R., McElvany, K.D. & Welch, M.J., 1982, Radiation Research, 90, 362.

Kassis, A.I., Fayad, F., Kinsey, B.M., Sastry, K.S.R., Taube, R.A. & Adelstein, S.J., 1987a, Radiation Research, 111, in the press.

Kassis, A.I., Nagasawa, H., Little, J.B. & Adelstein, S.J., 1986, In Abstracts of Papers for the Thirty-Fourth Annual Meeting of the Radiation Research Society, Las Vegas, Nevada (Radiation Research Society) p. 116.

Kassis, A.I., Sastry, K.S.R. & Adelstein, S.J., 1985a,

Radiation Protection Dosimetry, 13, 233.
Kassis, A.I., Sastry, K.S.R. & Adelstein, S.J., 1985b, Journal of Nuclear Medicine, 26, 59.
Kassis, A.I., Sastry, K.S.R. & Adelstein, S.J., 1987b, Radiation Research, 109, 78.
Kinsey, B., Kassis, A.I., Fayad, F., Layne, W.W. & Adelstein, S.J., 1987, Journal of Medicinal Chemistry, in the press.
Loevinger, R. & Berman, M., 1976, A Revised Schema for Calculating the Absorbed Dose from Biologically Distributed Radionuclides, MIRD Pamphlet No.1, revised (The Society of Nuclear Medicine).
Sastry, K.S.R., Howell, R.W., Rao, D.V., Mylavarapu, V.B., Kassis, A.I., Adelstein, S.J., Wright, H.A., Hamm, R.N. & Turner, J.E., 1987, In Proceedings of International Workshop on Auger Emitters in DNA: Implications and Applications, edited by K.F. Baverstock & D.E. Charlton (Taylor and Francis, Ltd.), p.27
Schneiderman, M.H. & Hofer, K.G., 1980, Radiation Research 84, 462.
Warters, R.L. & Hofer, K.G., 1977, Radiation Research, 69, 348.
Yasui, L.S. & Hofer, K.G., 1986, International Journal of Radiation Biology, 49, 601.
Yasui, L.S., Hofer, K.G. & Warters, R.L., 1985, Radiation Research, 102, 106.

INTERNAL AUGER EMITTERS: EFFECTS ON SPERMATOGENESIS
AND OOGENESIS IN MICE

D.V. Rao,[1] V.B. Mylavarapu,[1] K.S.R. Sastry,[2]
and R.W. Howell[2]

[1] Department of Radiology, University of Medicine &
Dentistry of New Jersey, Newark, NJ 07103, USA; and
[2] Department of Physics & Astronomy, University of
Massachusetts, Amherst, MA 01003, USA.

ABSTRACT

The in vivo biological effects of Auger emitters are investigated using (A) spermatogenesis in mouse testis, and (B) oogenesis in mouse ovary as experimental models. Sperm-head survival and induction of abnormal sperm, following intratesticular administration of radiopharmaceuticals, are the end points in Model A. Of interest in Model B is primary oocyte survival after intraperitoneal injection of the radiochemicals. The effectiveness of the Auger emitter is determined relative to its beta emitting companion or external X-rays in the absence of such an analogue. Our results reveal pronounced effects of Auger emitters on all the end points, and these do not depend on the mode of administration. The efficacy of the Auger emitter is related intimately to its subcellular distribution, which, in turn, is governed by the chemical form of the carrier molecule. Conventional dosimetry is inadequate and biophysically meaningful dosimetric approaches are needed to understand the in vivo effects of Auger emitters.

INTRODUCTION

The biological implications of internal Auger emitters are of much interest to basic science and to diagnostic and therapeutic nuclear medicine and radiation protection. These effects cannot be predicted a priori since they depend strongly on both the localisation of the radionuclides and the microscopic pattern of energy deposited by low energy Auger electrons in relation to intracellular radiosensitive targets, such as the nuclear DNA. The risks may be much more severe than expectations based on macroscopic considerations of conventional dosimetry (Loevinger and Berman, 1976). The likely importance of the effects and lack of experimental data motivated our studies on the in vivo toxicity of Auger emitters in experimental model systems.

EXPERIMENTAL MODELS
Model A. Spermatogenesis in mice

This basic process is complex but tractable (Meistrich et al., 1978). Amongst the several types of cells within the seminiferous tubules, the differentiated spermatogonia (Types A_1, A_2, A_3, A_4, In and B) are the most sensitive to ionising radiation. In contrast, their precursors as well as the postgonial cells (spermatocytes, spermatids, and spermatozoa) are relatively more radioresistant in the order given. This differential radiosensitivity is the basis for the study of radiation effects. When testicular cells are exposed to external X-rays or internal radionuclides, any lethal damage done to spermatogonia is manifested as reduced sperm-head population in the testis when these sperm heads are counted after the time necessary (4-5 wk) for spermatogonia to become spermatids (Meistrich et al., 1978). The sperm-head survival is assayed when the sperm-head count in the testis reaches a minimum. This optimal time is determined experimentally in each case. The mouse testis model is particularly valuable in studying the effects of Auger emitters in vivo because of its relevance to humans (Meistrich & Samuels, 1985).

At the end of the spermatogenic process, the spermatozoa migrate from the tubules to the epididymis, adjacent to the testis, and they are stored in this sac-like structure. (Meistrich et al., 1978.) The normal mouse sperm has a characteristic hook-like head. A variety of abnormalities of head-shape, stem, and tails are possible even in normal mice (Wyrobek, 1979; Bruce et al., 1974). Depending on the strain, a remarkably constant fraction of epididymal sperm is abnormal in normal mice. Environmental agents (radiation, chemicals, etc.) can elevate the abnormal sperm fraction above the spontaneous value. This induced abnormal fraction serves as a sensitive biological monitor of the effects of Auger emitters in the testes.

Model B. Oogenesis in the ovary

This is the process of the development of oogonia, the primordial germ cells in the ovary of the unborn mouse, into mature follicles leading to ovulation in the adult mouse. Soon after birth, the young mouse has 3,000 to 5,000 oocytes in the resting dictyate stage. Their number per ovary is limited and it cannot increase. Most of these cells are in the primary oocyte stage as small follicles;

Types 2 and 3a (Pederson & Peters, 1968). These are extremely radiosensitive with 50% survival at only 10 cGy X-ray dose (Oakberg & Clark, 1961). Accordingly, this system offers a very sensitive model to delineate the in vivo effects of Auger emitters.

MATERIALS AND METHODS

Male Swiss Webster mice (9-10 wk old), and 3-wk old females are used. Radiopharmaceuticals of interest are injected intratesticularly (i.t.) into males, and intraperitoneally (i.p.) into females. The advantages of i.t. mode in the delineation of effects of low energy electrons have been noted (Rao et al., 1983; 1985a). Since the ovaries are very small, the i.p. mode has been adopted in the case of female mice. The standard injection volume (3 µl i.t., and 100 µl i.p.) contains the radionuclides or chemicals of interest at the desired concentrations. Following the radiochemical injection, the pattern of elimination of the radionuclides from the organ is determined as a function of postinjection time. These results along with the radiation data are utilised to calculate the average cumulated radiation dose to the testis/ovary from varying amounts of injected radioactivity, using the MIRD schema based on conventional dosimetry (Loevinger & Berman, 1976).

Sperm heads are isolated and counted as described by Rao et al., (1983). Groups of about 50 mice are given i.t. injections of the same amount of the radiochemical, and the sperm-head count relative to unexposed control animals obtained on different days during the 6-7 wk postinjection period. The day when the relative sperm-head population attains a minimum (dip) is the optimal time for performing the sperm-head survival assay as a function of the testicular dose. The sperm produced in the tubules reach the epididymis 9 d later (Meistrich & Samuels, 1985). The optimal time for assay of shape abnormalities of epididymal sperm is, therefore, 9 d after the dip in sperm-head count is attained. At this time, the epididymal sperm are isolated, and slides are prepared using standard methods. The sperm are viewed under a light microscope (magnification 400) in bright field with a green filter. At least 2,000 are counted (normal + abnormal) and the ratio of abnormal sperm counts to the normal ones is obtained for several values of testicular doses. At different times after the i.t. injection, the macroscopic distribution of radionuclides in the testes is determined by measuring the activity per gram of testicular sections (Rao et al., 1985b). Frozen section autoradiography (Rao et al., 1983) is used to examine the microscopic distribution of the radionuclides in the organ. Cell fractionation methods are employed to determine the subcellular distribution of the radionuclides in vivo (Rao et al., 1985b). In the studies with external X-rays, the mice are shielded from whole body irradiation, so that the testes alone are exposed to X-rays. Possible chemotoxic effects of the nonradioactive chemicals are also investigated.

Seven days after i.p. injection of female mice, the primary oocycte survival is determined, relative to mice unexposed to the radiochemical, as a function of the dose to the ovary. The right ovary is isolated. Using standard histological procedures, 5 µm thick serial sections of the ovary are prepared. Using a light microscope with an oil immersion objective, the primary oocytes are counted in every 5th section to avoid recounting of the same. At least 25 to 50 such sections are scored for statistical accuracy.

As a standard approach, we have compared the effects of Auger emitters with their beta emitting counterparts, with similar distribution in the organ. Since beta rays have average ranges much greater than typical cell dimensions (~10 µm), they irradiate the cells and their nuclei quite uniformly irrespective of their cellular localisation. Accordingly, conventional macroscopic dosimetry should be quite valid in the case of beta emitters. The adequacy or inadequacy of the conventional approach in the case of Auger emitters should manifest itself in the relative dose-response curves for the two types of radionuclides. In the absence of suitable beta emitting companions, we compare the effects of Auger emitters with those of external X-rays.

RESULTS AND DISCUSSION

The radionuclide elimination from the testis and the ovary depends on the organ, mode of administration, the chemical nature of the radionuclide and the carrier. When injected via the i.t. mode as thallous chloride, the Auger/beta emitters, $^{201}Tl/^{204}Tl$ have the same clearance pattern despite their very different physical half-lives. The major (80%) and long-lived component has an effective half-life $T_e \sim$ 8-9 hr (Rao et al., 1983). For ^{99m}Tc Hydroxyate phosphate (^{99m}Tc-HDP) T_e is less than the physical half-life of ^{99m}Tc (6 hr). For such short values of T_e, all the dose is essentially delivered in the first 24hr. The dip in sperm-head count occurs on the 29th day postinjection as for single acute exposure to external X-rays (Figure 1, solid circles). For each of the indium radiochemicals (Rao et al., 1987), most of the initial activity is eliminated within minutes of i.t. injection, while a small fraction has a long biological half-life. In these cases, as well as ^{55}Fe/^{59}Fe-citrate and ^{75}Se and ^{35}S both in the form of methionine, values of T_e range from a few days to several days (Rao et al., 1985b, 1986, 1987). The dip occurs at about the 35th day postinjection (Figure 1, ^{111}In-oxine: solid triangles; ^{111}In-citrate: open circles). For such protracted exposure, the dose is calculated for the first 13-d period, consistent with the time scale of the spermatogenic process (see Rao, 1985b).

The radionuclide elimination from the ovary and the whole body have the same pattern. For ^{201}Tl/^{204}Tl, T_e = 18.0 and 29.8 hr, respectively, single exponential. For ^{111}In-oxine T_e = 43 hr, single exponential; and in the case of ^{111}In-citrate T_e = 7.2 and 54 hr for the 23% and 77% components, respectively. Unlike the spermatogonia, the primary oocytes are in a resting stage, their number cannot increase, and there is no regeneration following cytocidal effects. Essentially all the dose is delivered well before the 7th day, when primary oocyte survival is assayed.

Results of sperm-head survival studies are summarized in Table 1. Figures 2-5 are examples of the data from which the experimental values of relative biological effectiveness (RBE) are derived. The dose D_{37} at 37% survival is used to obtain RBE values in all cases. The two-component survival curve with external X-rays (Figure 3), and similar results of Gasinska (1985) in a different strain of mice, show that the observed two-component responses (Figure 2) with ^{201}Tl/^{204}Tl, ^{55}Fe/^{59}Fe (Rao et al., 1985b), ^{67}Ga-citrate (data not shown), and ^{75}Se/^{35}S-methionine (Rao et al., 1986) are not due to artifacts. The ^{111}In-oxine data are corrected for mild chemotoxicity of oxine (see Rao, 1987) and no synergism is observed between radiation and chemical (Rao et al., 1987). Although ^{114m}In-citrate is distributed similarly to ^{111}In-citrate, and ^{114m}In is an Auger emitter, it is the least toxic of the three indium radiochemicals. This is plausible since the dose is almost entirely (99.2%) from IC electrons (^{114m}In), and energetic beta rays (^{114m}In daughter). The RBE values of ^{201}Tl, ^{55}Fe, ^{111}In-oxine and ^{111}In-citrate qualitatively comparable. Even for ^{99m}Tc, the RBE is significantly greater than 1. The differences in the shapes of the survival curves as well as the RBE values most probably stem from the subcellular paterns of localisation of the Auger emitters and possible microdosimetric effects. ^{75}Se/^{35}S-methionine localise in the cytoplasm (Kassis et al., 1980), and the inefficacy of ^{75}Se

Figure 1. Time dependence of sperm-head population in mouse testes exposed to radiation. (Open circles - ^{111}In-citrate; solid circles - x-ray; solid triangles - ^{111}In-oxine).

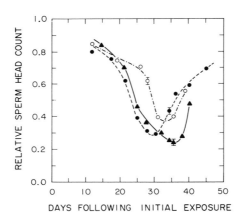

Figure 2. Sperm-head survival versus organ dose for ^{201}Tl and ^{204}Tl localised in the testis of mice.

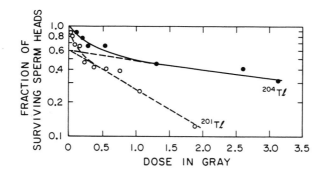

Effects on spermatogenesis and oogenesis 21

Figure 3. Sperm-head survival curve for external X-ray exposure of mouse testes. Solid (open) circles are for 60 (120) kVp X-rays.

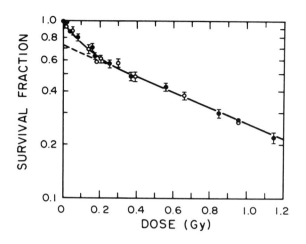

Figure 4. Sperm-head survival curves for ^{111}In-oxine (solid circles), ^{111}In-citrate (solid triangles) and ^{114m}In-citrate (open circles).

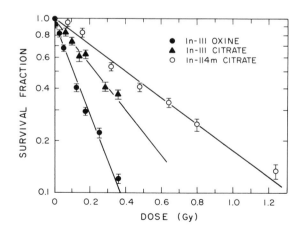

Figure 5. Sperm-head survival curve for ^{99m}Tc-HDP.

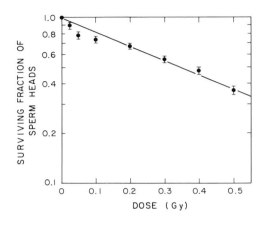

Figure 6. Abnormal sperm fraction versus testicular dose. Solid (open) circles are for ^{111}In-oxine (^{111}In-citrate); solid triangles are for X-rays.

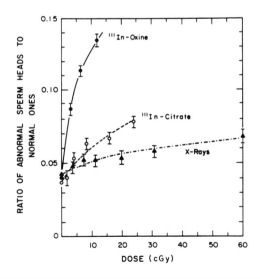

(RBE = 0.8) is not unexpected. Most interesting is the much stronger radiotoxicity of ^{111}In-oxine compared to ^{111}In-citrate. The RBE values (4.2 and 2.0, respectively) correlate approximately with the corresponding nuclear

Table 1. Auger emitters in mouse testes: Summary of results of sperm-head survival studies.

	Auger emitter and chemical form	RBE	Subcellular distribution
a.	^{201}TlCl	3.3*	Details unknown
b.	^{55}Fe citrate	2.6*	22% N; 78% Cy
c.	^{111}In oxine	4.2**	92% N; 8% Cy
c.	^{111}In citrate	2.0**	30% N; 70% Cy
c.	^{114m}In citrate	1.18**	30% N; 70% Cy
d.	^{75}Selenomethionine	0.8*	0% N; 100% Cy
e.	^{99m}Tc-HDP	1.4**	69% N; 31% Cy
e.	^{67}Ga citrate	1.6**	30% N; 70% Cy

*Relative to beta analogue. **Relative to X-rays. N = nucleus. Cy = Cytoplasm. (a. Rao et al., 1983; b. Rao et al., 1985b; c. Rao et al., 1987; d. Rao et al., 1986; e. Rao et al., (in press).

Figure 7. Percent of abnormal sperm as a function of testicular dose. Solid/open circles are for ^{201}Tl/^{204}Tl.

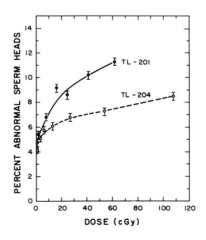

Figure 8. Primary oocyte survival versus dose to ovary. Solid/open circles are for ^{201}Tl/^{204}Tl.

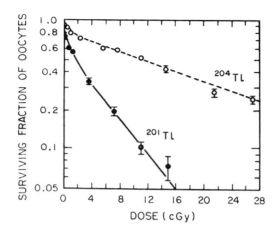

Figure 9. Primary oocyte survival versus dose to ovary from ^{111}In-oxine and ^{111}In-citrate.

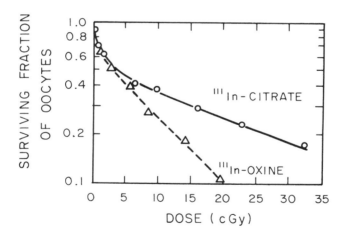

fractions (92% vs. 30%) of ^{111}In radioactivity (Table 1). These results point out the importance of the chemical nature of the carrier of the Auger emitter in the expression of biological effects in vivo (Rao et al., 1987).

Figures 6 and 7 show strong elevation of the abnormal to normal sperm ratio, over the spontaneous value, at very low doses. The initial slope (S) for ^{111}In-oxine is about 4 times the value for ^{111}In-citrate and 20 times the value for 120 kVp X-rays (Figure 6). The doubling dose (D) for ^{111}In-oxine is 2.5 cGy and for ^{111}In-citrate 25 cGy. For ^{201}Tl relative to ^{204}Tl, values of S are in the ratio of 2.5, and D = 15 cGy and 85 cGy, respectively. Whatever the criterion, these results are reminiscent of Wyrobek's (1979) findings with chemicals and provide the first evidence for the efficacy of Auger emitters in causing subtle damage to sperm.

Figure 8 shows that ^{201}Tl is far more efficient than ^{204}Tl in killing primary oocytes. The respective D_{37} values are 3.5 cGy and 17.5 cGy. The RBE of ^{201}Tl relative to ^{204}Tl is 5.0, compared to 3.3 with sperm-head survival as the end point. Data in Figure 9 give a value of 1.7 for ^{111}In-oxine relative to ^{111}In-citrate, the higher efficacy stemming from the chemical differences between the two.

In conclusion, this work reiterates our initial finding of the effectiveness of Auger emitters in causing biological effects in vivo (Rao et al., 1983). Enhanced effects are seen with three different end points, independent of the mode of administration of the radiochemicals. The toxic effects of Auger cascades depend on the intracellular localisation and subcellular distribution of the Auger emitter, the chemical form of the agent being the determinant. Macroscopic treatment of Auger electrons without regard to the above is responsible for the inadequacy of conventional dosimetry. The biological dosimetry presented here should provide a phenomenological basis for improved dosimetry of Auger emitters.

ACKNOWLEDGEMENTS

It is a pleasure to acknowledge the contributions of V.K. Lanka, G.F. Govelitz and H.E. Grimmond to this research. This work is supported by US PHS Grant CA-32877.

REFERENCES

Bruce, W.R., Furrer, R. & Wyrobek, A.J., 1974, Mutation Research, 23, 381.
Gasinska, A., 1985, Neoplasma, 32, 443.
Kassis, A.I., Adelstein, S.J., Haydock, C. & Sastry, K.S.R., 1980, Radiation Research, 84, 407.
Loevinger, R. & Berman, M., 1976, Medical Internal Radiation Dose Committee Pamphlet No. 1, Revised (Society of Nuclear Medicine, New York).
Meistrich, M.L., Hunter, N.R., Suzuki, N., Trostle, P.K. & Withers, H.R., 1978, Radiation Research, 74, 349.
Meistrich, M.L. & Samuels, R.C., 1985, Radiation Research,

102, 138.
Oakberg, E.F. & Clark, E., 1961, Journal of Cell and Comparative Physiology, 58, 173.
Pedersen, T. & Peters, H., 1968, Journal of Reproduction and Fertility, 17, 555.
Rao, D.V., Govelitz, G.F. & Sastry, K.S.R., 1983, Journal of Nuclear Medicine, 24, 145.
Rao, D.V., Sastry, K.S.R., Govelitz, G.F., Grimmond, H.E. & Hill, H.Z., 1985a, Radiation Protection Dosimetry, 13, 245.
Rao, D.V., Sastry, K.S.R., Govelitz, G.F., Grimmond, H.E. & Hill, H.Z., 1985b, Journal of Nuclear Medicine, 26, 1456.
Rao, D.V., Govelitz, G.F., Sastry, K.S.R. & Howell, R.W., 1986, In Proceedings of 4th International Radiopharmaceutical Dosimetry Symposium, CONF 85-1113, edited by A.T. Schlafke-Stelson & E.E. Watson, p.52.
Rao, D.V., Sastry, K.S.R., Grimmond, H.E., Howell, R.W., Govelitz, G.F., Lanka, V.K. & Mylavarapu, V.B., 1987, Journal of Nuclear Medicine (in press).
Wyrobek, A.J., 1979, Genetics, 92, 105.

DOSIMETRY OF AUGER EMITTERS: PHYSICAL AND PHENOMENOLOGICAL APPROACHES

K.S.R. Sastry,[1] R.W. Howell,[1] D.V. Rao,[2]
V.B. Mylavarapu,[2] A.I. Kassis,[3] S.J. Adelstein,[3]
H.A. Wright,[4] R.N. Hamm,[4] and J.E. Turner[4]

[1]Department of Physics & Astronomy, University of Massachusetts, Amherst, MA 01003, USA;
[2]Department of Radiology, University of Medicine & Dentistry of New Jersey, Newark, NJ 07103, USA;
[3]Shields Warren Radiation Laboratory, Harvard Medical School, Boston, MA 02115, USA; and
[4]Oak Ridge National Laboratory, Oak Ridge, TN 37831, USA

ABSTRACT

Recent radiobiological studies have demonstrated that Auger cascades can cause severe biological damage contrary to expectations based on conventional dosimetry. Several determinants govern these effects, including the nature of the Auger electron spectrum; localised energy deposition; cellular geometry; chemical form of the carrier; cellular localisation, concentration, and subcellular distribution of the radionuclide. Conventional dosimetry is inadequate in that these considerations are ignored. Our results provide the basis for biophysical approaches toward subcellular dosimetry of Auger emitters <u>in vitro</u> and <u>in vivo</u>.

INTRODUCTION

The MIRD Schema (Loevinger & Berman,1976) is widely used at present to estimate the biological risks of internal radionuclides. This conventional dosimetric approach assumes that radionuclides and the radiation energy are uniformly distributed in organs. Even though the cell is the basic unit at which biological effects are expressed, the calculated average dose to the organ is tacitly presumed to be the dose to the cell and its nucleus as well. The quality of radiation, the nature of the radiochemical, possible cellular concentration of radionuclides, and their subcellular distribution are ignored. In spite of this, conventional dosimetry has served as a reasonable approximation for energetic electrons, beta rays, and photons with ranges and mean free paths in biological matter greater than several cell diameters. In the case of Auger emitters, the above assumptions and inadequacies may severely restrict the validity of the conventional approach. It is true that

Auger electrons carry only a small fraction of the energy released per decay, and contribute negligibly to the total organ dose. However, by spreading out their energy over macroscopic dimensions, conventional dosimetry trivializes the likely importance of these low energy Auger electrons.

The biological implications of Auger emitters have been of interest for nearly 20 years. The severe toxicity of DNA incorporated ^{125}I has been demonstrated repeatedly in various cell systems. Our in vitro and in vivo studies (Kassis et al., 1987; Rao et al., 1987) show that Auger emitters cause biological damage depending on their localisation relative to the cell, its nucleus, and the DNA which is the primary target of radiation action. The biological dosimetry and biophysical results emerging from these studies provide an experimental data base for meaningful approaches to the dosimetry of Auger emitters at the subcellular level. We examine some of the physical and phenomenological aspects in this paper.

PHYSICAL CONSIDERATIONS
Auger cascades

Photoelectric effect and nuclear decay by orbital electron capture (EC) and internal conversion (IC) cause inner shell ionisation of atoms. The complex atomic vacancy cascades that follow are dominated by nonradiative (NR) Auger, Coster-Kronig (CK) and super CK processes illustrated in Figure 1 (Sastry & Rao, 1984). In each NR transition filling a vacancy, two new vacancies are created in the higher atomic subshells. This vacancy multiplication, and the progressively decreasing transition energies result in a dense shower of low energy electrons (~20 to 500 eV) with very short ranges (~1 to 20 nm) in biological matter. The review by Sastry & Rao, (1984) of the physics and biophysical dosimetry of Auger cascades contains tables of average Auger and CK electron spectra calculated for several radionuclides of interest. Relevant references may be found therein. Figure 2 indicates the complexity of the electron spectrum. The results shown for ^{111}In decay are obtained using Monte Carlo methods (Charlton & Booz, 1981; Howell et al., 1986) assuming no appreciable charge build-up on the ^{111}Cd daughter in the condensed phase. The electron groups at the extreme right (Figure 2) are due to IC. Table 1 is a summary of average Auger and CK electron yields for several radionuclides. Electrons with very low energies (< 500 eV) are preponderant in each case. The Auger yields depend on the atomic number, EC and IC probabilities, and the number of steps in isomeric decays (e.g., ^{193m}Pt & ^{195m}Pt). In a

Figure 1. Nonradiative atomic vacancy transitions.

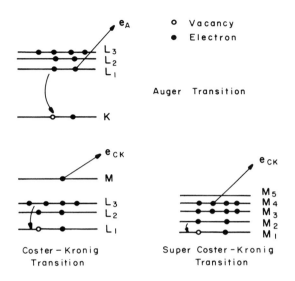

Figure 2. Electron spectrum for ^{111}In decay.

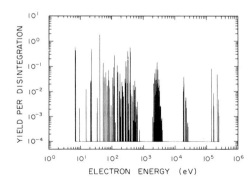

single decay, the electron spectrum depends on the relaxation pathways, and it can be very different from the average spectrum.

Localised energy deposition: High LET effects

Densely ionising α-particles, with their high linear energy transfer (LET) of about 100 keV/μm in unit density matter, are very efficient in inducing biological effects. The low energy Auger electrons have LET values ~15 to 25 keV/μm (Cole, 1969). Simultaneous ejection of several cascade electrons should be expected to simulate high LET-type effects in the immediate vicinity of the decay site. Using the methods of Kassis et al., (1980), we have calculated the average energy ($\triangle E$) deposited by the electrons in a unit density sphere, 5 nm in radius around the decay site. These results, obtained in the continuously slowing down approximation (csda), are given in Table 1. Possible contributions from potential energy on the ion are not included. Figure 3 shows the energy absorbed in concentric spherical shells of various radii, centered on the Auger emitter (^{193m}Pt, ^{111}In & ^{55}Fe). These curves (Wright et al., 1987) are the results of Monte Carlo calculations for liquid water using the Oak Ridge Electron Transport Code. The dashed curve for ^{193m}Pt, obtained using the average spectrum and csda, compares reasonably with the Monte Carlo results. This illustrates the utility of simple and approximate methods, especially in view of basic uncertainties in atomic transition rates and energies in the near-valence region. The differential profile of energy density around the Auger emitters ^{125}I and ^{193m}Pt is shown (Figure 4) by the dashed and solid curves, respectively. These results indicate the pattern of the highly localised energy deposition (HILED), the energy density dropping precipitously by two orders of magnitude in the first 5 nm. Considering that 10 eV/(nm)^3 implies a local dose equivalent of 1.6 MGy in commonly understood terms, the data bring to light the implications of small energy depositions in extremely minute volumes. Figure 5 shows that the high density of chemical species (dots) around the Auger emitter ^{193m}Pt in liquid water is comparable with the density of species along the track of a 4-MeV alpha particle (Wright et al., 1987).

BIOPHYSICAL ASPECTS
Positional and concentration effects

Values of $\triangle E$ in Table 1, the patterns of HILED (Figures 3 & 4), and the high density of chemical species in the immediate vicinity of the decay site of the Auger emitter

Figure 3. Localised energy deposition. '———' Monte Carlo calculations for 500 decays '— — — —' c.s.d.a with average electron spectrum.

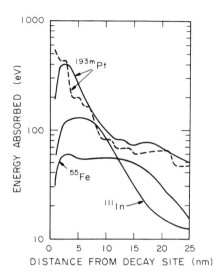

Table 1. Average yields of Auger and CK electrons and localised energy deposition for some radionuclides.

Radionuclide	Average yield per decay*	Energy (eV) deposited in a 5 nm sphere
^{51}Cr	6 (5)	210
^{55}Fe	5 (3)	240
^{67}Ga	5 (3)	260
^{75}Se	7 (5)	270
^{77}Br	7 (5)	300
^{99m}Tc	4 (4)	280
^{111}In	8 (7)	450
^{125}I	19 (17)	1000
^{193m}Pt	27 (23)	1800
^{195m}Pt	33 (27)	2000
^{201}Tl	20 (17)	1400

*Low energy electron yield is given in parentheses.

Figure 4. Profile of absorbed energy density calculated using c.s.d.a. '___', ^{193m}Pt, '_ _ _', ^{125}I.

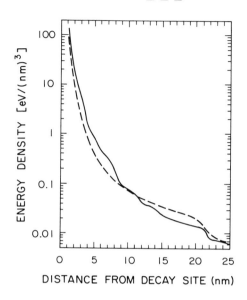

Figure 5. Track structure in liquid water for a 4-MeV alpha particle and ^{193m}Pt Auger cascades.

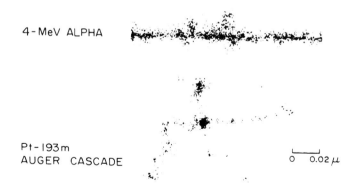

(Figure 5) indicate the potential for severe damage of molecules in the neighbourhood. The in vitro studies (Kassis et al., 1987) show that ^{125}I and ^{77}Br, covalently bound to the DNA of Chinese hamster V79 cells, kill the cells efficiently, the relative biological effectiveness (RBE) being 6-7 (Kassis et al., 1987). The RBE of ^{125}I attached to a DNA intercalator is about 5 for the same cell line. The drug trans-platin (II) forms adducts with the DNA. Our recent work shows that, when cells are exposed to this platinum-coordinated complex labelled with ^{195m}Pt, the RBE of this Auger emitter is also about 5 (Kassis et al., 1987). In contrast, ^{125}I when almost entirely (96%) in the cytoplasm is not so efficient (RBE ~1.3). Decays of ^{125}I and ^{75}Se in the cytoplasm are mildly radiotoxic, while extracellular decays of ^{125}I and ^{77}Br are much less so. The radiotoxicity of ^{51}Cr stems from the fact that 15% of the activity in the cell is in its nucleus. The in vivo effects also depend on the localisation of the Auger emitter (Rao et al., 1987). When ^{75}Se remains in the cytoplasm of testicular cells and in the intertubular spaces, the RBE is essentially unity. For ^{111}In-oxine, with 92% of the cellular radioactivity in the nucleus, the RBE is about 4 (Rao et al., 1987). The RBE is much lower for ^{111}In-citrate with only 30% of intracellular ^{111}In present in the nucleus. The effects of low energy electrons should depend on the location of the Auger emitter. These studies confirm such a dependence and quantify the effects as a function of the location of the radionuclide.

The experimental studies clearly reveal the importance of the chemical nature of the carrier in determining the subcellular distribution of the radionuclide. Cells in vitro concentrate some radionuclides greatly (Kassis et al. 1985, 1987). Relative to extracellular concentration, ^{125}IdU concentration in the nuclei of V79 cells is 1850-fold larger, after an 18-hr incubation at 37^{0}C. For ^{75}Se-methionine, the cellular concentration factor n = 330, and for ^{201}TlCl, n = 130. The ^{201}Tl studies, in fact, demonstrate that cellular localisation of this Auger emitter is essential for manifestation of its biological effects (Kassis et al., 1983). The in vivo studies cannot easily establish evidence for concentration of radionuclides by testicular cells. Nevertheless, biochemical mechanisms in vivo and in vitro should be reasonably similar, and cellular concentration of radionuclides should be one of the determinants of the effects of Auger emitters.

Subcellular energy deposition

The above considerations point to the importance of the energy deposited in the target cell and its nucleus from decays occurring within the cell. Kassis et al., (1980) have calculated the energy deposited in the cytoplasm and in the nucleus of V79 cells from decays in the respective compartments assuming that the nucleus and the cell are concentric spheres of unit density (Figure 6). Hypothetical radionuclides, emitting monoenergetic electrons with unit yield, are assumed to be distributed uniformly in the cytoplasm (Cy) or the nucleus (N) of the cell. Values of energy deposited - ε_{NN}, ε_{NCy} and ε_{CyCy} - are calculated as a function of the electron energy per decay. The first subscript represents the target region, and the second one, the source region. The results do not depend strongly on deviation from spherical geometry but do depend on the nuclear radius R_N and the cell radius R_C. Figure 7 shows the results for cells with R_C = 4.5 µm and R_N = 2.5 µm, the average geometry of spermatogonial cells in our mice.

REALISTIC DOSIMETRY
Cells in culture

In these studies, suspended cells are kept separated by shaking during incubation. The dose to the target cell nucleus is the sum of the dose from decays external to the cell (E_{Ne}) and from internal decays (E_{NN} and E_{NCY}). The total amount of energy deposited in the nucleus $E_N = E_{NN} + E_{NCy} + E_{Ne} = d_t\varepsilon_N$. The contribution of E_{Ne}, calculated using MIRD Schema, is negligible. The quantity $\varepsilon_N = r_N \varepsilon_{NN} + r_{Cy} \varepsilon_{NCy}$, where r_N and r_{Cy} are, respectively, the nuclear and cytoplasmic fractions of intracellular radioactivity. The cumulated number of decays in the cell, d_t, is the sum of the disintegrations occurring during incubation (d_I), and postincubation (d_{PI}). These are experimentally determined from uptake data during incubation, and from postincubation retention studies, both as a function of time. Values of ε_{NN} and ε_{NCy} are obtained for each radionuclide using its electron spectrum and the generalised energy deposition curves similar to those in Figure 7. For ^{125}I, ε_{NN} = 11.2 keV/decay and ε_{NCy} = 0.86 keV/decay. The vast difference between the two shows the high efficacy of decays in the nucleus compared to those in the cytoplasm.

Spermatogonial cells: Dose enhancement

The actual dose received by the nuclei of cells should

Figure 6. Geometry for calculation of subcellular energy deposition. See Kassis et al., (1980) for details.

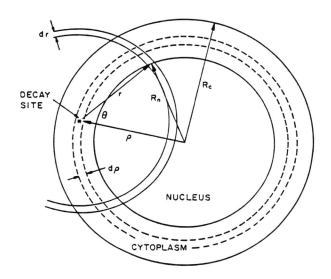

Figure 7. Generalised energy deposition curves for spermatogonial cells: Curves A, B and C represent ε_{NCy}, ε_{CyCy} and ε_{NN}, respectively. See text for full explanation.

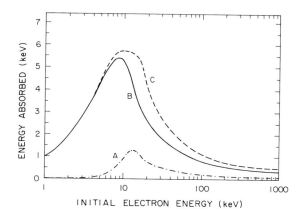

be larger than conventional estimates in view of possible concentration of the Auger emitter by cells. For cells in culture, all the variables are generally known and the cumulated dose is readily calculated. This is not so for spermatogonial cells <u>in vivo</u>. We have, therefore, introduced a dose enhancement factor (N_N) (Rao et al., 1985). Here $N = R_N/R_{CON}$, R_N being the actual instantaneous rate of energy deposition in the target nucleus from decays within the cell and from the external region, and R_{CON}, the conventionally calculated value. The following expression for N_N is readily derived:

$$N_N = \{(n/f_N)(r_N\Phi_{NN} + r_{Cy}\Phi_{NCy})/[f_c(n-1) + 1] + \Phi_{Ne} + \triangle\}(1 + \triangle)^{-1}, \qquad (1)$$

where f_c = fraction of the testicular volume occupied by the spermatogonial cells, and f_N = fraction of the spermatogonial cell volume occupied by its nucleus. The quantities Φ_{NN} and Φ_{NCy} are, respectively, the ratios of electron energy absorbed in the target nucleus to the electron energy emitted per decay in the nucleus, and in the cytoplasm of the target cell. The parameter Φ_{Ne} is the ratio of electron energy absorbed in the target nucleus to the electron energy emitted per decay external to the target cell. The symbol $\triangle$ is the ratio of photon energy absorbed in the testis to the electron energy emitted per decay in the organ. The variables n, r_N and r_{Cy} are defined earlier.

In equation (1), Φ_{Ne} and $\triangle$ are essentially given by conventional dosimetry. For the measured cellular geometry, f_N = 0.171. The absorbed fractions Φ_{NN} and Φ_{NCy} are determined using the results in Figure 7 and the electron spectra for various Auger emitters. The subcellular distributions (r_N and r_{Cy}) are measured. Thus, n and f_c are the free variables. For spermatogonia in the 10 μm basal layer of the tubules (diameter 100 μm), a value of $f_c \sim 0.1$ is a reasonable one (Rao et al., 1985). Values of N_N are predicted consistent with the observed RBE values (Rao et al. 1987) for reasonable values of n. Examples of the the parameters are given in Table 2. The respective experimental RBE values for ^{55}Fe-citrate, ^{111}In-oxine and ^{111}In-citrate are 2.6, 4.3 and 2.0. Equation (1) predicts these values for N_N with n = 5 (^{55}Fe), n = 12 (^{111}In-oxine), and n = 10 (^{111}In-citrate).

CONCLUSION

In this paper, we have examined the nature of HILED

Table 2. Parameters in the dose enhancement factor.

Parameter	^{55}Fe-citrate	^{111}In-oxine	^{111}In-citrate
r_N	0.15	0.92	0.30
r_{Cy}	0.85	0.08	0.70
Φ_{NN}	0.89	0.122	0.122
Φ_{NCy}	0.016	0.0048	0.0048
Φ_{Ne}	0.0	0.892	0.892
$\triangle$	0.332	0.114	0.114

primarily responsible for the high LET-type effects of Auger emitters close to or incorporated into DNA. The possible role of potential energy on the ion needs further investigation. The prevalence of highly localised doses around the site of the emitter should be expected to cause a variety of severe lesions in the DNA. This feature may have implications for radiotherapy and public health. We have found that calculation of doses to the nucleus is reasonable even when Auger emitters are DNA-bound. The dose enhancement factors (N_N) are based on our phenomenological results, and the parameters are identified. The expression for N_N adequately explains the observed RBE values in the mouse testis. The results of primary oocyte studies are, in principle, understandable along similar lines. Our model for N_N is also relevant to the in vitro situation where $f_c \sim 0$. At 37% survival, we estimate a value of about 1,000 for N_N in the case of V79 cells incubated with ^{125}IdU for 18 hr. This illustrates the inadequacy of conventional dosimetry for situations in vitro. When f_c has values of about 0.7 for highly packed cell systems, $N_N \rightarrow 1$ even for large values of n. In this limit, as well as when n → 0, conventional dosimetry is essentially valid.

This work is supported in part by USPHS Grants CA 32877 and CA 15523, and by USDOE Contract DE-AC05-84OR21400.

REFERENCES
Charlton, D.E. & Booz, J., 1981, Radiation Research, 87,

10.
Cole, A., 1969, Radiation Research, 38, 7.
Howell, R.W., Sastry, K.S.R., Hill, H.Z. & Rao, D.V., 1986, In Proceedings of 4th International Radiopharmaceutical Dosimetry Symposium, CONF 85-1113, edited by A.T. Schlafke-Stelson & E.E. Watson, p. 493.
Kassis, A.I., Adelstein, S.J., Haydock, C. & Sastry, K.S.R., 1980, Radiation Research, 1984, 407.
Kassis, A.I., Adelstein, S.J., Haydock, C. & Sastry, K.S.R., 1983, Journal of Nuclear Medicine, 24, 1164.
Kassis, A.I., Sastry, K.S.R. & Adelstein, S.J., 1985, Radiation Protection Dosimetry, 13, 233.
Kassis, A.I., Adelstein, S.J., Howell, R.W. & Sastry, K.S.R., 1987, In Proceedings of International Workshop on Auger Emitters in DNA: Implications and Applications, edited by K.F. Baverstock & D.E. Charlton (Taylor & Francis) p.1
Loevinger, R. & Berman, M., 1976, Medical Internal Radiation Dose Committee Pamphlet No. 1, Revised (Society of Nuclear Medicine).
Rao, D.V., Sastry, K.S.R., Govelitz, G.F., Grimmond, H.E. & Hill, H.Z., 1985, Journal of Nuclear Medicine, 26, 1456.
Rao, D.V., Mylavarapu, V.B., Sastry, K.S.R. & Howell, R.W., 1987, In Proceedings of International Workshop on Auger Emitters in DNA: Implications and Applications, edited by K.F. Baverstock & D.E. Charlton (Taylor and Francis) p.15
Sastry, K.S.R. & Rao, D.V., 1984, In Physics of Nuclear Medicine: Recent Advances, edited by D.V. Rao, R. Chandra & M. Graham (Medical Physics Monograph No. 10) (American Institute of Physics), p. 169.
Wright, H.A., Hamm, R.N., Turner, J.E., Howell, R.W., Sastry, K.S.R., Rao D.V. & Haydock, C., 1987, Abstracts: Thirty-fifth Annual Meeting of the Radiation Research Society, Abstract Cr-3, p. 48.

HIGH-LET ENDORADIOTHERAPY: IN VIVO STUDIES WITH A
METABOLICALLY-DIRECTED AUGER EMITTING ANTICANCER DRUG
IN A MURINE TUMOUR MODEL

I. Brown, R.N. Carpenter & J.S. Mitchell[†]

The Research Laboratories, The Radiotherapeutic Centre,
Cambridge University School of Clinical Medicine,
Addenbrooke's Hospital, Cambridge CB2 2QQ, U.K.

INTRODUCTION

The development of metabolically directed radioactive anti-cancer agents, which selectively target towards neoplastic cells, offers an alternative approach towards the management of locally advanced and metastatic malignant disease (Brown, 1982a; Mitchell et al., 1983; Brown, 1987). Implicit in this approach is not only the requirement that suitable compounds exhibit a degree of selective concentration within viable tumour cells, preferably stem cells, but also the stable incorporation of an appropriate radionuclide in the carrier molecule. By using high specific activity compounds, it may be possible to exert an appreciable _in situ_ radiotherapeutic effect on cancer cells, with minimal damage to normal tissue.

Optimal endoradiotherapy would be best achieved by using compounds that stably incorporate radionuclides which decay by the emission of radiations of a high-LET quality, as they possess many advantages over the low-LET radiations commonly employed in clinical radiotherapy (Fowler, 1985). The relative biological effectiveness (RBE) of ionizing radiations has been observed to rise steadily with increasing LET, reaching a maximum at $\sim$100 keV/μm, above which higher energy deposition quantitatively results in a cellular overkill-effect (Hall, 1978). _In vitro_ cell studies have also shown that the cytotoxicity attributed to high-LET radiations is independent of dose rate and cell cycle (Sasaki, 1984); cell damage is lethal, being predominantly due to non-rejoining DNA double strand breaks (Ritter et al., 1977). There is no cellular repair of

[†] Sadly, Professor J.S. Mitchell FRS died during the latter stages of this work.

sublethal damage or potentially lethal damage after high-LET irradiation (Barendsen, 1964; Bertsche et al., 1983). Furthermore, the oxygen enhancement ratio (OER) observed for high-LET radiations is near unity (Barendsen et al., 1966; Schöpfer et al., 1984), so enabling comparable cytotoxicities to be effected in both euoxic and hypoxic tumour cell sub-populations.

The radionuclide ^{125}I with a half-life of 60.2d appears to have very suitable properties. It is an abundant emitter of Auger and Coster-Kronig electrons due to its complex electron capture and internal conversion decay modes (Charlton & Booz, 1981). Monte-Carlo calculations have indicated that the number of electrons per decay varies between 1 and 56 with an average value of 21.2 in the condensed phase (Charlton et al., 1978); the great variation in the total number of electrons emitted per decay and total energy carried by these electrons (~700 eV to 70 keV, mean value ~19 keV) depends upon the de-excitation route of the daughter tellurium atom (Charlton, 1986).

It has been well established that iodine-125 can exert severe biological damage at both cellular as well as molecular levels; when incorporated into the DNA of proliferating cells via the thymidine analogue, 5-[^{125}I]-iododeoxyuridine. The high toxicity of its Auger emissions lead to biological effects usually associated with those due to high-LET particles (Hofer et al., 1975; Chan et al., 1976; Krisch et al., 1977), which is consistent with the microdosimetric concept of higher energy deposition in small critical targets. In contrast, its cytotoxicity is minimal and of a low-LET character when it decays in the cytoplasm, cell membrane or at extracellular locations (Kassis et al., 1985). The RBE of Auger electrons has been estimated as about 4-8 (Brown et al., 1982).

Hofer (1980) in his treatment of ^{125}I dosimetry indicated that ^{125}I decaying at the centre of small spheres deposits more energy than that of a 5 MeV α-particle traversing the same diameter (<50nm), and so concluded that biochemically bound ^{125}I can decay producing high-LET effects. A more recent microdosimetric study by Charlton (1986) of ^{125}I localized at the central axis of a DNA duplex and at distances from the axis, has demonstrated a sharp fall off in energy deposition approaching 4nm from the decay point. It would appear that high-LET damage only occurs within 1-2nm of DNA, damage being predominantly of low-LET character at larger dimensions. These conclusions are generally in accord with experimental results in biological systems, apart from those of Commerford et al.,

(1980) who reported high-LET effects due to ^{125}I in the cell nucleus but not bound to DNA, though under their experimental conditions the intimate proximity of DNA and ^{125}I cannot be discounted.

Clearly, to exert a high-LET advantage ^{125}I must be incorporated into the nuclear genome, such that it is bound or in close proximity to DNA. This essential prerequisite represents an obvious limitation to the implementation of Auger emitting compounds as endoradiotherapeutic agents as it necessitates their specific sequestration to intra-nuclear elements.

In vivo therapeutic studies using an early murine ovarian ascites model have shown that 5-[^{125}I]-iodo-2-deoxyuridine exhibits marked antineoplastic activity (Bloomer & Adelstein, 1977); this was attributed to the nuclear toxicity of ^{125}I incorporated into DNA. Similarly, Feinendegen et al., (1986) reported encouraging results (31% partial response) in mice bearing transplanted sarcoma-180, after a 10d infusion of 7.4 MBq ^{125}IdU. These results prompted the limited clinical investigation of 5-^{125}IdU into two patients with advanced oropharyngeal carcinoma; 3.7 - 4.1 x 10^7 Bq ^{125}IdU was arterially infused over 12 - 44d. There was no evidence of bone marrow toxicity; tumour response was poor (Feinendegen et al., 1986; Boerker, private communication).

Presently, a novel class of metabolically-directed high-LET radiohalogenated potential endoradiotherapeutic drugs is being studied (Brown, 1982a; Mitchell et al., 1983b). Substituted 1,4-naphthoquinol diphosphates (MNDP) are known to be selectively taken up by some malignant tumours both in vitro (Marrian et al., 1961) and in vivo (Marrian et al., 1969). The mechanism of tumour localization depends upon the presence of oncogenically associated membrane alkaline phosphatase isoenzymes of placental sub-group origin, followed by sequestration of active metabolite into nuclear structures (Brown, 1987; Carpenter & Brown, 1987).

One of these compounds, 6-[^{125}I]-iodo-MNDP (Figure 1) has been studied in in vitro cell systems (Brown et al., 1982; Carpenter et al., 1983) and in vivo animal studies (Brown et al., 1987; Brown & Carpenter, 1987). In vitro clonogenic cell survival-dose studies have been undertaken with cultured human epidermoid carcinoma of the larynx (HEp2) cells (Toolan, 1954) and murine rectal adenocarcinoma (CMT-93) cells (Franks & Hemmings, 1978), exposed to 6-[^{125}I]-iodo-MNDP, under both euoxic and hypoxic conditions. Cell survival-dose curves were resolvable as single negative exponential curves, characteristic of high-LET radiations, D_o values were

1.38 and 1.68 mBq/cell, and 1.44 and 1.76 mBq/cell respectively; the OER was <1.5 (Carpenter et al., 1983; Brown, 1987). Autoradiography has confirmed the intranuclear localization of 6-[^{125}I]-iodo-MNDP in both HEp2 and CMT-93 cells.

Biodistribution studies in CMT-93 rectal tumour bearing mice of the two radiohalogenic homologues, 6-[^{125}I]-iodo-MNDP and 6-[^{211}At]-astato-MNDP have demonstrated tumour selectively and metabolic stability of the respective covalent carbon-radiohalogen bonds (Brown et al., 1984). In order to evaluate the potential therapeutic value of 6-[^{125}I]-iodo-MNDP as an in vivo high-LET endoradiotherapeutic drug, the efficacy of its associated Auger and Coster-Kronig emissions have been studied in the murine CMT-93 rectal tumour model.

Figure 1. 6-[^{125}-I]-iodo-2-methyl-1, 4-naphthoquinol diphosphate (6-[^{125}I]-iodo-MNDP)

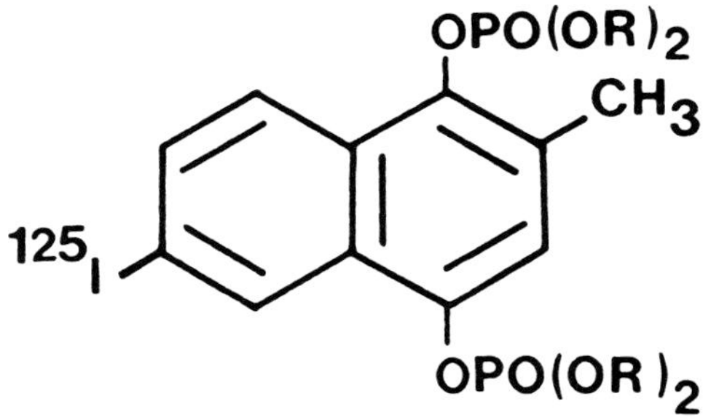

METHODS

Radiochemistry

High specific activity 6-[^{125}I]-iodo-MNDP (238 MBq/μmol) was prepared by thermal homogeneous isotopic exchange with 6-iodo-MNDP under in vacuo conditions (Brown, 1982b). The product was separated by ion-exchange chromatography and purified by high pressure liquid chromatography.

Sodium [^{125}I]-iodide (7.8x10^{4}MBq) was obtained from commercial sources (Amersham International PLC).

Both compounds were prepared as buffered (phosphate-free) aqueous solutions to pH 7.4, and sterilized by membrane filtration (Millipore; 0.22 μm) before injection.

Dose aliquots were determined by the measurement of 27-32 keV Te K X-rays using a 2" NaI (Tl) crystal well counter.

Therapy
Therapeutic studies with both 6-[^{125}I]-iodo-MNDP and [^{125}I]-iodide anion were undertaken in male C57BL10 mice of mean weight 22.3 $\pm$ 1.5g, each of which possessed a single well delineated subcutaneous transplanted, moderately well differentiated rectal adenocarcinoma in its flank. Mice were generally housed five to a cage, with food and water provided ad libitum.

Mice were treated when the tumour volume was in the exponential growth range, at approximately $10^2 - 10^3$ mm^3, when the mean doubling time was 3.2 $\pm$ 0.5d. Their general physical condition was monitored by clinical examination which included whole body weight determinations, initially as with tumour volume measurements (on the basis of three maximal perpendicular dimensions), on a daily basis but after 3-4 weeks at longer intervals. Death was taken as the end point of each therapeutic trial; however, mice were sacrificed if deemed to be in extremis.

Therapy involved the single intraperitoneal injection of either 6-[^{125}I]-iodo-MNDP or sodium [^{125}I]-iodide solutions; doses ranged from 0.925-22.2 MBq. One hour prior to treatment, each mouse received a subcutaneous injection of potassium perchlorate (0.25mg/kg) in order to block thyroid function, and so minimise the uptake of free [^{125}I]-iodide into the thyroid gland.

Wherever possible, all deceased animals were kept and temporarily stored at -30^{0}C until a detailed necroscopic examination could be performed in order to ascertain the cause of death.

RESULTS AND DISCUSSION
Mice were observed to tolerate [^{125}I]-endoradiotherapy well, over the whole dose range studied. There was no clinical evidence of any acute radiation induced syndrome or anaphylactic reaction. The carrier of concentration of 6-iodo-MNDP was well below that associated with any pharmacological toxicity.

Survival for at least six months, with a negligible 'residual tumour' volume of $\leq$10 mm^3 without any clinical evidence of metastases was regarded as satisfying the criteria for a 'cure'. Untreated control tumour-bearing mice, all inevitably died due to either the local or systemic consequences of tumour progression, generally within five weeks after tumour transplantation.

The therapeutic response of the CMT-93 rectal tumour to

Figure 2. The effect of a single intraperitoneal injection of 6-[^{125}I]-iodo-MNDP on tumour growth in cured and non-cured CMT-93 rectal tumour bearing mice.

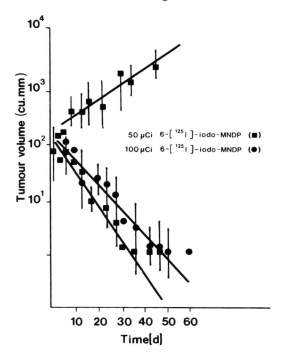

6-[^{125}I]-iodo-MNDP was observed in an early change in tumour growth kinetics, (Figure 2). The prolongation of tumour doubling time in 'non-cured' mice was found to be dependent upon the administered radioactive dose of 6-[^{125}I]-iodo-MNDP.

The survival fraction (Φ_n) of treated animals after a single intraperitoneal injection of 6-[^{125}I]-iodo-MNDP could be related to the administered radiation dose D (activity) by a Langmuir-type saturation curve of the form

$$\Phi_n = \frac{\xi_n D}{(1 + \zeta_n D)} \qquad (1)$$

at n months where ξ_n and ζ_n are constants. These equations are generally applicable to kinetic processes in which specific receptor sites are implicated (Benson, 1960), and are consistent with the substrate role of 6-[^{125}I]-iodo-

Figure 3. The survival fraction (Φ) of CMT-93 rectal tumour-bearing mice at 6 and 12 months after a single intraperitoneal injection of 0.93-22.2MBq 6-[^{125}I]iodo-MNDP.

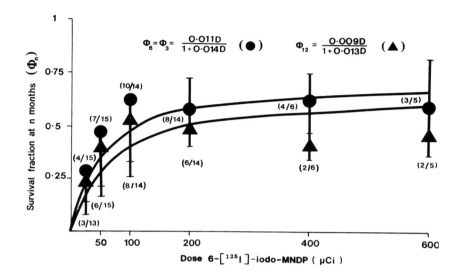

MNDP for oncogenically-associated alkaline phosphatase isoenzymes (Brown, 1987).

Survival-dose data arising from this model have been statistically fitted to equation (1). Typical survival fraction (Φ_n)-dose (D) curves are shown in Figure 3; these demonstrate a plateau therapeutic effect of 45-55% at a dose of 7.4 MBq 6-[^{125}I]-iodo-MNDP.

If the survival fraction at 6 months (Φ_6) is considered representative (see Figure 2), then over the complete dose range of 6-[^{125}I]-iodo-MNDP the crude survival rate 59.3% (32/54) was significantly different ($\chi^2 = 657$; $P \ll 10^{-4}$) from that of untreated control mice (1/232; 0.4%). The analysis of comparable survival fractions (Φ_n) relative to different time intervals (n) did not reveal any significant differences; tumour cures appeared to be permanent.

In an attempt to quantify the possible tumourcidal radiation dose delivered by 6-[^{125}I]-iodo-MNDP, semi-empirical radiodosimetric calculations based upon kinetic data derived from 6-[^{125}I]-iodo-MNDP biodistribution studies (Brown, 1987; Brown & Carpenter, 1987) and from parallel quantitative intracellular α-autoradiographic analyses for the higher homologue, 6-[^{211}At]-astato-MNDP (Mitchell et al., 1983b, 1984; Brown, 1987) were undertaken. If a CMT-93 tumour bearing mouse weighing 25g received 7.4 MBq 6-[^{125}I]-iodo-MNDP, it was estimated that the mean electron dose delivered to tumour cell nuclei would be approximately 50 cGy.[1] The contribution from the photon dose to the nucleus would be ~19 cGy. It was assumed that all the Auger electrons within the cell nucleus exhibited a high-LET behaviour, and so the RBE was taken as at least 5. Consequently, it was assumed that the radiobiologically equivalent single dose of ^{60}Co γ-rays corresponding to the uniform distribution of 6-[^{125}I]-iodo-MNDP metabolite in tumour cell nuclei would be ~270 cGy. The distribution of 6-[^{125}I]-iodo-MNDP in the CMT-93 tumour is heterogeneous. It was also assumed that the conclusions drawn from in vivo α-autoradiographic studies for 6-[^{211}At]-astato-MNDP parallel those for 6-[^{125}I]-iodo-MNDP, the proportion of cells which may be considered of stem cell or clonogenic character is ~7-10% (Brown, 1987; Brown et al., 1987). Approximately 40% of these cells have α-particles originating from within the nuclear envelope, excluding the nuclear membrane. It follows that the radiobiologically equivalent single dose of ^{60}Co γ-rays to the whole tumour would be in the order of 61 - 114 Gy.[2] Though it is recognised that by its nature, this type of calculation can incur significant errors, doses of this magnitude are in accord with experimental evidence and clinical experience. Therapeutic studies have established that in the murine CMT-93 rectal tumour model: a) 7.4-22.2MBq 6-[^{125}I]-iodo-MNDP lead to tumour cures in 45-60% animals, whilst b) single fraction 6 MeV electron teletherapy over the range 35-55 Gy was not curative (Brown, 1987).

Similar calculations have been made for critical normal tissues (Table 1). Estimated radiation doses were modest, and were clearly influenced by the homogeneous distribution of 6-[^{125}I]-iodo-MNDP metabolite within these tissues, and the lower proportion of ^{125}I loci within intranuclear structures. Even so these values are probably over-estimates. The relatively higher absorbed dose ascribed to the lungs is due to its higher RBE for high-LET radiations. There was no histological evidence of late radiation damage

---

[1]Mean Biological Concentration = 2.1 (Brown & Carpenter, 1987)

[2]Full details of this calculation are given by Brown (1987)

Table 1. Biologically-equivalent low-LET single dose estimations for 6-[^{125}I]-iodo-MNDP in some critical normal tissues of mice[1] (Brown, 1987)

Tissue	Absorbed radiation dose (cGy)[2]		Mean proportion[3,4] intranuclear loci of ^{125}I	Approximate low-LET equivalent[5] dose (cGy)
	Intranuclear	Cytoplasmic		
Lungs	4.83	7.97	0.53 ± 0.12	31 – 61[6]
Spleen	2.11	3.49	0.55 ± 0.38	2 – 8
Liver	6.64	10.96	0.66 ± 0.31	11 – 55
Testis	4.52	7.48	0.70 ± 0.12	11 – 35
Bone Marrow	1.81	2.99	0.66 ± 0.10	5 – 13
Colon	3.92	6.48	0.55 ± 0.11	9 – 26

1. For a 25g CMT-93 tumour-bearing mouse that had received 200 μCi (7.4 MBq) 6-[^{125}I]-iodo-MNDP.
2. Relative energy deposition (intranuclear:cytoplasm = 23.34 : 38.56).
3. Including nuclear membrane.
4. Not corrected; a higher proportion of intranuclear localization at 0.5 – 6 h.
5. RBE of Auger/Coster-Kronig electrons within cell nuclei (not including nuclear membrane) was taken as 15 for pulmonary tissue; and 5 for other tissues (compared with ^{60}Co γ-rays).
6. Example calculation: [(4.83 ± 0.40) × 15 × (0.53 ± 0.12)] + (7.97 ± 0.71) = 46.4 ± 14.6 cGy.

in tissue from long-term survivors.

Although inorganic [^{125}I]-iodide was well tolerated by all mice, there was no evidence of any regression or delay in tumour growth. There was no significant difference in survival between CMT-93 tumour bearing control mice and those that received sodium [^{125}I]-iodide.

CONCLUSION

Therapeutic studies with high specific activity 6-[^{125}I]-iodo-MNDP have demonstrated that this Auger-emitting drug is effective in treating mice with a transplanted CMT-93 rectal adenocarcinoma without incurring early or late radiation-associated sequelae. Its efficacy has been attributed to the selective intranuclear localization of 6-[^{125}I]-iodo-MNDP metabolite and consequent high-LET quality toxicity in heterogeneously distributed tumour cells that exhibit specific onco-alkaline phosphatase isoenzyme activity. Quasi-microdosimetric estimations of the radiobiologically equivalent low-LET single fraction absorbed tumour dose are of an order of magnitude that might be associated with the permanent control of this tumour.

ACKNOWLEDGEMENTS

We gratefully acknowledge financial support from the Professor Joseph Mitchell Cancer Research Fund (R.N.C.) and Beit Memorial Trust (I.B.). We are indebted to Mary-Ann Starkey for her contribution to final preparation of this manuscript.

REFERENCES

Barendsen, G.W., Koot, C.J., Van Kersen, C.R., Bewley, D.K., Field, S.B. & Parnell, G.J., 1966, International Journal of Radiation Biology, 10, 317.

Benson, S.W., 1960, The Foundation of Chemical Kinetics, McGraw-Hill Book Co. Inc., New York, Toronto, London, pp. 703.

Bertsche, U., Iliakis, G. & Kraft, G., 1983, Radiation Research, 95, 57.

Brown, I., 1982a, In Nuclear Medicine and Biology, Vol. 1, Raynaud, C., (ed), Pergamon Press, Paris, 166.

Brown, I., 1982b, Radiochemical Radioanalytical Letters, 52, 283.

Brown, I., 1987, High Linear Energy Transfer Endoradiotherapeutic Drugs for Malignant Disease, MD Thesis, University of Cambridge, UK., pp. 437.

Brown, I. & Carpenter, R.N., 1987, Nuklearmedizin (in press).

Brown, I., Carpenter, R.N. & Mitchell, J.S., 1987, International Journal of Radiation Oncology, Biology and Physics (in press).
Brown, I., Carpenter, R.N. & Mitchell, J.S., 1982, European Journal of Nuclear Medicine, 7, 115.
Carpenter, R.N., Brown, I., 1987, Biochemical Journal (in press).
Carpenter, R.N., Brown, I. & Mitchell, J.S., 1983, International Journal of Radiation Oncology, Biology and Physics, 9, 51.
Charlton, D.E., 1986, Radiation Research, 107, 163.
Charlton, D.E. & Booz, J., 1981, Radiation Research, 87, 10.
Charlton, D.E., Booz, J., Fidorra, J., Smith Th. & Feinendegen, L.E., 1978, In Sixth Symposium on Microdosimetry, Booz, J. & Ebert, H.G. (eds), Harwood Academic Publishers Ltd, for the Commission of the European Communities, 91.
Commerford, S.L., Bond, V.P., Cronkite, E.P. & Reincke, U., 1980, International Journal of Radiation Biology, 37, 547.
Fowler, J.F., 1985, International Journal of Radiation Biology, 47, 115.
Franks, L.M. & Hemmings, G.T., 1978, Journal of Pathology, 124, 35.
Hall, E.J., 1978, Radiobiology for the Radiologist, Harper and Row, London, pp. 460.
Hofer, K.G., 1980, Bulletin Cancer (Paris), 67, 343.
Kassis, A.I., Sastry, K.S.R. & Adelstein, S.J., 1985, Radiation Protection and Dosimetry, 13, 233.
Krisch, R.E., Krasin, F. & Sauri, S.J., 1977, Current Topics on Radiation Research Quarterly, 12, 355.
Mitchell, J.S., Brown, I. & Carpenter, R.N., 1983a, Experientia, 39, 337.
Mitchell, J.S., Brown, I. & Carpenter, R.N., 1983b, International Journal of Radiation Oncology, Biology and Physics, 9, 57.
Ritter, M.A., Cleaver, J.W. & Tobias, C.A., 1977, Nature, 266, 653.
Sasaki, H., 1984, Radiation Research, 99, 311.
Schöpfer, K., Schneider, E., Rase, S., Kieffer, J., Kraft, G. & Liesem, H., 1984, International Journal of Radiation Biology, 46, 305.
Toolan, H.W., 1954, Cancer Research, 14, 660.

RADIOCHEMOTHERAPY WITH ^{125}I-5-IODO-2-DEOXYURIDINE

K.D. Bagshawe,
Cancer Research Campaign Laboratories,
Department of Medical Oncology,
Charing Cross Hospital,
LONDON. W6 8RF U.K.

Cytotoxic drugs have been in clinical use now for 40 years and it has become apparent that, with the exception of a group of comparatively uncommon tumours, drug resistance is an almost universal response. Whilst holding the view that we will not fully comprehend what we call drug resistance without understanding why a few cancers can be eliminated with these drugs, there can be no doubt that cytotoxic agents tend to select out cell clones resistant to their own action.

When cytotoxic drugs are given to patients with drug resistant cancer, the cancer continues to grow but toxic effects on normal tissues are always incurred. If the cytotoxic agents are inhibitors of DNA synthesis then this would seem to imply that DNA synthesis continues in the tumour whilst it is arrested in normal cell renewal tissues. Surprisingly, there does not appear to have been any formal demonstration that this dissociation of DNA synthesis occurs until our own recent studies (Bagshawe, 1986; Bagshawe et al., 1987). If this dissociation is a general phenomenon then it appears to represent a gross difference between cancerous and normal tissues which has been neglected.

If DNA synthesis continues in S phase cancer cells when it is temporarily inhibited in normal tissues the question arises whether we can introduce some lethal agent into the cancer cells. Since the blockade of DNA synthesis in normal cells can only be safely maintained for some hours the lethal agent should have a short half life. 5-iodo-2-deoxyuridine (IdU) is simply thymidine with an iodine molecule replacing the 5-methyl group of the thymidine. It has been known for many years as a radiosensitiser and ^{125}IdU has been used as a substitute for ^{3}H thymidine in laboratory studies as a marker for DNA synthesis because it can compete with and substitute for thymidine in the formation of DNA. In the late 1950's it was shown that because of the decay characteristics of ^{125}I, i.e. producing an Auger electron cascade, ^{125}IdU is a highly toxic substance (Prusoff, 1959).

When IdU is injected in vivo some is incorporated into the DNA of S-phase cells where it appears to remain until the cell dies or passes it to the daughter cells. IdU not incorporated into DNA is rapidly metabolised with a 1/2 life in man of the order of 3-4 minutes and about 7 minutes in the mouse.

When ^{125}IdU is incorporated into DNA its decays produce apparently irreparable double strand breaks with most of the energy of the disintegration being dissipated within 20-25 nm of the site of decay. As the distance between DNA and decay of ^{125}I increases toxicity diminishes. Hofer, (1980) found that about 500 disintegrations produced a 99% kill of Chinese hamster ovary cells when ^{125}I was in the form of ^{125}IdU and thus in association with the nucleus but when it was attached to the cell membrane as ^{125}I concanavalin A about 100 times more disintegrations were required per cell to achieve the same toxicity.

For comparative purposes it is interesting that about 25,000 molecules of ^{125}IdU would produce some 500 disintegrations in each of the daughter cells in 4 days and that the rate of incorporation of thymidine molecules into human DNA in an S phase of 10 hours duration is of the order of 20,000 per second. Also, in broad terms ^{125}I has 5-7 times the cytotoxic efficiency of ^{131}I in DNA or of x-rays.

The block to using ^{125}IdU therapeutically came when it was found that injection into tumour bearing animals led to uptake by normal tissues, particularly by intestinal tract where uptake was many times greater than that by even rapidly growing tumours. So it has, it seems, remained for 30 years a sort of "if only" agent - if only we could get into tumours selectively.

In the laboratory either hydroxyurea (HU) or cytosine arabinoside are used to arrest DNA synthesis. Whilst these agents are used clinically, mainly for certain leukaemias they are relatively ineffective against most solid cancers, many of which appear resistant to them de novo or quickly acquire resistance. Our studies so far have mainly used HU which inhibits ribonucleotide reductase. In our human xenograft studies we have used CC3, a choriocarcinoma which is resistant to HU and LS174T (kindly donated by Schlomm) a moderately differentiated adenocarcinoma of colon which is only partially resistant to HU.

In addition to taking advantage of the sensitivity of normal tissues to HU and the resistance of tumours to HU we have attempted to augment cancer cell uptake of ^{125}IdU by using fluoropyrimidines to block the enzyme thymidylate synthetase and anti-folates to block the folic acid mechanism which provides the CH_3 group necessary for synthesis of thymidine. When thymidine synthesis is blocked, cells can use the thymidine salvage pathway utilising thymidine or IdU present in extra cellular fluid.

Summarising a substantial number of experiments carried out in the last 18 months we can conclude as follows:

1. When ^{125}IdU was injected IV alone into nude mice carrying the CC3 tumour, the tumour to small intestine ratio (T:S1) was 0.11 by counting tissues removed after killing the animal at 24 hours.

2. When HU 50 mg/kg was given 5 minutes before the injection of ^{125}IdU the T:S1 was 4.4. For all other tissues the ratio was more favourable.

3. Using various sequences of methotrexate, 5-fluorouracil, hydroxyurea and ^{125}IdU. T/S1 ratios of 12:1 with the CC3 tumour and 20:1 with the LS174T tumour have been obtained. The highest % retention of administered dose was 0.7% for CC3 and 1.5% for LS174T.

4. Area under the curve studies over a 7 day period gave a calculated dose for the LS174T/SI of 13:1.

5. Repeated doses under hydroxyurea blockade of normal cells resulted in dose accumulation in the CC3 tumour.

6. Dose escalation over the range 0.37-37 MBq ^{125}IdU showed increase in uptake with the higher doses but with some fall in % of total dose retained. Studies in which cold IdU was added also indicated that saturation of uptake was not attained in the dosages used.

7. Autoradiographic studies showed that at 24 hours ^{125}IdU was predominantly located over tumour cell nuclei. With the CC3 model T:S1 ratios by grain counting in some experiments were >100:1.

We have labelled IdU with ^{131}Iodine and investigated this as a scanning agent in patients using 370-555 MBq ^{131}IdU with hydroxyurea, methotrexate and folinic acid. Good images have been obtained in some but not all cases. Broadly speaking, they are of similar quality to imaging with radiolabelled antibody but gastric secretion of free ^{131}I is marked in some patients.

Pilot therapeutic studies have been started in patients with advanced drug resistant disease using 1000-2000 MBq doses of ^{125}IdU. Myelosuppression has occurred but it is not yet clear whether the ^{125}IdU contributes to this or whether it results entirely from the concomitant chemotherapy. The biopsies obtained so far suggest satisfactory exclusion of ^{125}IdU from normal cell nuclei. It is too early to attempt any assessment of therapeutic response.

In conclusion drug resistance allows us to arrest DNA synthesis in normal renewal populations and to introduce a potentially lethal nucleotide analogue, ^{125}IdU, selectively into tumour cell DNA. Clinical investigation will require the development of a range of new monitoring techniques especially for serum thymidine.

Acknowledgements

These studies have been supported by the Cancer Research Campaign. I wish to thank my collaborators, Surinder Sharma, Dr. Frances Searle, Joan Boden, Robert Boden, Dr. Barbara Pedley, Simon Riggs, Geoffrey Boxer and Dr. Peter Southall.

References

Bagshawe, K.D., 1986, The Lancet, 778-780.
Bagshawe, K.D., Boden, J., Boxer, G.M., Britton, D.W., Green, A., Partridge, T., Pedley, B., Sharma, S., Southall, P., 1987, British Journal of Cancer 55, 299-302.
Hofer, K.G., 1980 In: Third International Radiodosimetry Symposium: US Dept. of Health and Human Services p. 371. Health and Human Services p.371.
Prusoff, W.H., 1959, Biochemica & Biophysica Acta 32, 295.

DNA DAMAGE BY AUGER EMITTERS

Roger F. Martin†, Barry J. Allen*, Glenn d'Cunha†, Richard Gibbs†, Vincent Murray† and Marshall Pardee†.

†Molecular Science Group, Peter MacCallum Cancer Institute, 481 Little Lonsdale Street, Melbourne

*Lucas Heights Research Laboratories, Australian Nuclear Science and Technology Organisation, Lucas Heights, N.S.W., Australia

ABSTRACT

The interest in DNA damage by Auger emitters is largely motivated by the crucial significance of DNA as a radiobiological target. More specifically, such studies contribute to our understanding of the well-established biological effects of Auger emitters, typified by the classical ^{125}IdU suicide experiments. However, DNA also provides a special opportunity to analyse more fundamental aspects of the radiochemical damage induced by decay of Auger emitters. For example, ^{125}I atoms can be introduced at specific locations along a defined DNA target molecule, either by site-directed incorporation of an ^{125}I-labelled deoxynucleotide or by binding of an ^{125}I-labelled sequence-selective DNA ligand. After allowing accumulation of ^{125}I decay-induced damage to the DNA, application of DNA sequencing techniques enables the positions of strand breaks to be located relative to the site of decay, at a resolution corresponding to the distance between adjacent nucleotides (0.34 nm). In other words, DNA provides a molecular framework upon which to analyse the extent of damage following (averaged) individual decay events. The results can be compared with energy deposition data generated by computer-simulation methods developed by Charlton and co-workers. The DNA sequencing technique also provides some information about the chemical nature of the termini of the DNA chains produced following Auger decay-induced damage.

In addition to reviewing the application of this approach to the analysis of ^{125}I decay induced DNA damage, some more recent results obtained by using ^{67}Ga will also be presented.

INTRODUCTION

In the early 1920's Pierre Auger was using a Wilson cloud chamber to investigate the photoelectric effect and noticed short tracks at the start of many of the photoelectron tracks. The length of the short tracks was independent of the energy of the incident X-rays, but did vary between the different gases irradiated. Auger (1925) correctly interpreted the origin of the electrons as transitions between orbitals in the process of filling the vacancy created by the photoelectron. Some of cloud chamber photos showed multiple Auger electron tracks from a single ionisation for which Auger later used the description 'electronic strip-tease' (Auger, 1975).

The prerequisite for Auger electron emission is a vacant inner shell orbital, so the Auger effect is not confined to photoionisation; electron capture and internal conversion are two examples of processes which also lead to the Auger electron cascade. The Auger emitting isotope that has received most attention is ^{125}I which decays by successive electron capture and internal conversion. It is beyond the scope of this paper to discuss the ^{125}I decay process in detail, for that the reader is referred to Charlton and Booz, (1981) and other chapters in this book.

Although the Auger effect had been implicated by many early workers interpreting the radiobiological effects of incorporated ^{125}I, it was Schmidt and Hotz, (1973) who first provided a molecular basis for those effects. By using neutral sucrose gradient sedimentation to measure double-strand (ds) DNA breaks in T1 phage, they established the 1:1 relationship between ^{125}I decay and ds DNA breaks, which was confirmed in later studies (Krisch & Ley, 1974; Krisch & Sauri, 1975; Krisch, Krasin & Sauri, 1976). The importance of ds DNA breaks had been suggested earlier by Krisch (1972). It seems surprising that the ds DNA break assay using electrophoresis of circular ds phages or plasmids, which is much more accurate than sedimentation of larger phages, has not been used to verify the 1:1 dogma. The theoretical calculations of Charlton (this volume) may prompt such a re-assessment.

The purpose of this paper is to describe results that continue the trend set by the early measurements of ds DNA break induction in ^{125}I-labelled DNA, and which attempt to provide a molecular basis to link the Auger effect and biological end-points.

METHODS

The details of all the methods used are contained in the references cited in the Figure legends and/or the text.

One exception is the neutron irradiations relevant to Figure 6. The source of thermal neutrons was the 88 KW Moata reactor at Lucas Heights with a thermal neutron column described by Allen et al., (1986) which at full power generates a flux of 10^{10} neutrons cm^{-2} sec^{-1}. All samples for irradiation contained 5 µg pBR322 in 100 µl buffer containing 10 mM each of Hepes and Tris HCl at pH 7.5, except where indicated otherwise.

RESULTS AND DISCUSSION
^{125}I-labelled DNA ligands induce ds DNA breaks

The potential of Auger emitters in cancer therapy, highlighted by Bloomer and Adelstein, (1977), had been appreciated for some time, but the ^{125}IdU 'suicide' approach had obvious limitations. The labelled DNA precursor can not discriminate between different cell populations, and is restricted to cells in the S-phase of the cell cycle. The question of whether ^{125}I that is non-covalently associated with DNA can induce a ds DNA break was significant. Its implications became apparent when Bloomer et al., (1980) showed that ^{125}I-labelled tamoxifen was cytotoxic, and when the induction of ds DNA breaks by ^{125}I-triiodothyronine was reported by Sundell-Bergman and Johanson, (1982). Both these studies demonstrated the possibility of exploiting receptor-mediated endocytosis for delivery of Auger emitters to the DNA of particular cell populations, namely those with appropriate receptors.

Induction of ds DNA breaks by an ^{125}I-labelled DNA ligand was first demonstrated with an ^{125}I-labelled aminoacridine (Martin, 1977) using ds circular PM2 DNA as an assay substrate. A similar ^{125}I-labelled acridine was later shown to be cytotoxic, presumably due to ds DNA break induction (Martin, Bradley and Hodgson, 1979). However the significance of ds DNA break induction by non-covalently associated ^{125}I was not restricted to the implications for a cytotoxic modality.

The result suggested that the Auger effect could be used to study sequence-selectivity of DNA ligands by application of DNA sequencing techniques (discussed later) and this 'proximity probe' concept has been exploited in a study of the interaction between ^{125}I-α-Bungarotoxin and its receptors (Schmidt, 1984). Also, from the radiobiological standpoint, the fact that an ^{125}I-labelled DNA ligand can induce ds DNA breaks enables studies of ds DNA break induction at different phases of the cell cycle.

Finally, the result makes some contribution to the debate, crystallised by Hofer, Keogh and Smith (1977), as to whether molecular fragmentation or electron irradiation is the dominant mechanism for ds DNA break induction. The elegant study of Deutzmann and Stocklin (1981) clearly established ring fragmentation as a consequence of the Auger effect in ^{125}I-iodouracil. However the extent to which ring fragmentation may contribute to ds DNA break induction is not clear. Although positive charges originating on the pyrimidine ring might be transferred along the polynucleotide chain and induce breaks (Linz and Stocklin, 1985), it is more difficult to envisage a similar process originating from positive charges on a DNA ligand.

DNA sequencing techniques - DNA as a defined framework

In Figure 1 the protocol for analysing DNA damage using DNA sequencing techniques is given. The deoxyribose units are the radiosensitive moiety of the DNA double duplex where radiochemical attack results in strand scission (von Sonntag et al., 1981).

The application of this approach to ^{125}I damage in DNA is illustrated in Figure 2, reproduced from the study of Martin and Haseltine (1981), which showed that the 'averaged' ^{125}I decay induces single-strand (ss) breaks in both near (^{125}I containing) and opposite strands of the DNA duplex. The majority of strand breaks occur within 4 to 5 bp of the deoxycytidine that was labelled with ^{125}I.

The same method has been used to study the strand breakage induced by ^{125}I-iodoHoechst 33258, a sequence selective DNA ligand (Martin and Holmes, 1983). More recently we have compared the binding sites detected by the Auger effect with those determined by footprinting (Dervan, 1986), a typical result is shown in Figure 3. Note that the binding site includes four consecutive AT base pairs, a consensus sequence common to virtually all Hoechst 33258 binding sites. The displacement of the hot-spot of damage from the ^{125}I-ligand from the footprint position demonstrates the bias introduced in the former method. This bias suggests that some end-labelled ^{32}P-DNA strands sustain more than one break as a result of ^{125}I decay, since only the break nearest the end-label will be 'scored'. This distribution of strand break frequencies indicated by the intensities of the bands in the hot-spot is similar to that described for decay of covalently bound ^{125}I (Martin and Haseltine, 1981), which is in accord with recent theoretical calculations by Charlton (this volume).

Taken together, the DNA sequencing gel data suggest that the ^{125}I induced ds DNA break results from single-strand breaks in both strands. Moreover in view of the observed 'bias' effect, individual decay events sometimes induce more than one ss break in either (or both) strand(s), so

Figure 1. Analysis of damage using DNA sequencing techniques. A defined fragment of ds DNA end-labelled with ^{32}P is depicted at the top of the diagram. The degree of strand breakage at each deoxyribose group reflects the radiation dose at that site. Open circles represent deoxyribose units. Electrophoresis of the resulting fragments on a DNA sequencing gel fractionates single-strand chains according to size. Intact target DNA will have the lowest mobility (left); shorter strands move further. Unlabelled DNA chains derived from the lower strand or from the left-hand side of breaks in the upper strand are not shown; only ^{32}P-end-labelled strands are detected by autoradiography. The intensity of each band reflects the degree of breakage at the corresponding site. Appropriate markers are used to correlate bands on the autoradiograph with particular sites in the nucleotide sequence of the initial DNA fragment.

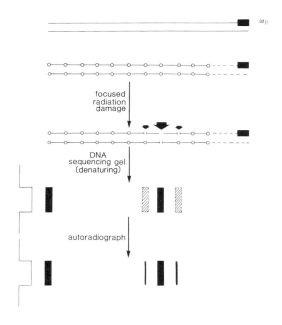

the final lesion may not simply be a ds DNA break, but a 'mini deletion', due to loss of short fragments at the site of breakage, at least in some cases.

All the preceding experiments have been done <u>in vitro</u> using purified end-labelled DNA. How relevant are the results to Auger electron-induced DNA damage in intact cells, with all the complexities of chromatin structure?

Figure 2. Analysis of damage resulting from decay of ^{125}I covalently bound to a defined site in the opposite strand (reproduced from Martin and Haseltine, 1981). The species depicted at the right hand side of the diagram was obtained by annealing two single-strand fragments of plasmid DNA, one containing the ^{125}I and the other the ^{32}P end-label. The nucleotide sequence is aligned with the data from the markers in lanes 1 to 4 (G, G+A, C+T and C, respectively). Incubation of the hybrid for 44 days results in the appearance of fragments corresponding to breakage opposite the ^{125}I (lane 5), compared to a relatively fresh sample of the hybrid. Densitometric analysis of the portion of autoradiograph was used to estimate the range of damage relative to the site of ^{125}I decay.

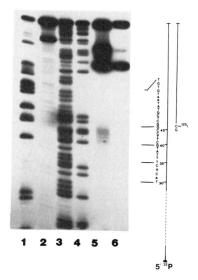

^{125}I-induced ds DNA breaks in intact cells

We have exploited the sequence selectivity of binding/damage by ^{125}I-Hoechst 33258 to study ds DNA breaks induced in intact cells. Methods have been described in detail (Murray and Martin, 1985a, 1985b, 1985c) but two features should be noted. Firstly the studies employ a 'target' sequence in human cells called αRI-DNA. The 340 bp repeat unit is present in about 100,000 copies in the human genome, and most importantly, the sequence homogeneity amongst the multiple copies is very high - about 91%. Secondly, in order to study ds DNA damage, we have used αRI-DNA M13 clones (single-stranded) as hybridisation reagents to 'fish-out' αRI-DNA fragments from non-αRI-DNA.

Figure 3. Damage by ^{125}I-labelled iodoHoechst 33258. The 375 bp EcoRI-Bam HI fragment from pBR322 DNA, 3'-end-labelled with ^{32}P was incubated alone (C-left) or with 3.6 x 10^7 cpm of carrier-free ^{125}I-iodoHoechst 33258 (T) and incubated in a buffer comprised of 20 mM Tris, 50 mM NaCl, 1 mM EDTA and 1 mM KI) (final concentrations) for 22 days at 5°C. Other samples of the ^{32}P-labelled DNA were pre-incubated in dH$_2$O with either no addition (C) or with cold iodoHoechst 33258 at final concentrations of 5 µM (T$_5$) or 10 µM (T$_{10}$), for 30 min at 25°C. An equal volume buffer comprising 10 mM NaCl$_2$, 10 mM MgCl$_2$, 1 mM MnCl$_2$ and 5 mM CaCl$_2$ (final concentrations) containing DNAase I at 3.8 µg/ml (C), 3.8 µg/ml (T$_5$), 38 µg/ml (T$_{10}$) and incubation continued for 1 min before termination with excess EDTA. The sequencing reactions in the two lefthand lanes indicate that the Hoechst binding site detected by ^{125}I iodoHoechst cleavage (T) and by DNAase I "footprinting" (lanes T$_5$ and T$_{10}$) is within the sequence AATTTAA.

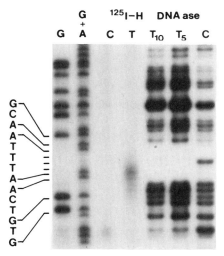

The outstanding feature of the results, illustrated in Figure 4, is the marked similarity between damage induced in vitro with purified DNA, and that after exposure of intact cells to ^{125}I-Hoechst 33258. There are several implications of this result, but for the present discussion the most important one is that the 'mini-deletion' type lesion is also likely to be a feature of Auger damage in intact cells.

In an attempt to find further evidence for the 'mini deletion', we began a study of ^{125}I-induced mutants at the HPRT locus in CHO cells, using classical ^{125}IdU suicide. The potential of such a study is that DNA sequencing of the HPRT mutants could reveal the nature of the ultimate mutagenic lesion (a mini deletion?). The results were surprising (Gibbs et al., 1987). Although the number of mutants examined was low, the majority of ^{125}I-mutants 5) of 6) had substantial deletions, easily detected

---

Figure 4. Comparison of damage induced by ^{125}I-iodoHoechst 33258 in purified genomic DNA <u>in vitro</u> vs. intact human K562 cells. Approximately 6×10^7 cpm of carrier-free ^{125}I-iodoHoechst 33258 was incubated for 1 hr at 20°C in 50 µl of a buffer containing 20 mM Tris pH 7.9, 110 mM NaCl, 5% mM EDTA, 5 mM KI and 10% glycerol (v/v), and including 2×10^6 K562 cells (lane 2+3), 10^6 K562 cells (lane 4) or 10 µG (lane 5+6) of purified high molecular weight DNA from K562 cells. Lanes 2 and 5 did not contain ^{125}I-iodoHoechst 33258. Lane 1 is a Maxam-Gilbert G+A sequencing track. The samples were then stored for 31 days at -70°C to accumulate ^{125}I-decay-induced damage. Details of the subsequent procedures for DNA isolation, ^{32}P-labelling, purification of α-DNA and analysis have been described (Murray and Martin, 1985 a,c).

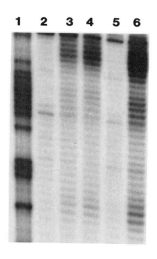

by Southern analysis. Moreover in two mutants the deletions were so large (more than 20-30 Kbp) that no HPRT coding sequences could be detected by Southern analysis. The nature of the lesion in the mutant with an undamaged Southern pattern (i.e., the only likely candidate for a mini deletion) has not been established.

A possible interpretation of the results is that mini deletions, which can only be properly repaired by a recombination type mechanism, sometimes ultimately become very much larger deletions, as a result of an unsuccessful attempt at recombination type repair.

^{67}Ga-labelled DNA ligands

In looking toward the possible therapeutic utility of DNA ligands labelled with Auger emitting isotopes, there are a number of reasons why ^{125}I is a poor candidate. Most of the contra-indications relate to radioprotection considerations; long half-life, easy dehalogenation of many iodo-compounds and consequent release of ^{125}I vapour, and the low permissible body burden due to efficient concentration by the thyroid. We have therefore begun a study of alternative Auger emitters, which could be coupled to DNA ligands, with most of our attention focussed on ^{67}Ga.

For several reasons we chose a desferrioxamine as the chelating moiety, and it has been coupled to an aminoacridine and to a bisbenzimidazole derivative. ^{67}Ga is then introduced to yield the final conjugates. Most of our experiments to date with these compounds have been concerned with ds DNA break induction using the circular plasmid DNA assay. The typical experiment involves incubating the ^{67}Ga-desferrioxamine DNA ligand conjugate with pBR322 DNA and then using agarose gel electrophoresis to detect linear species. A radiation control of ^{67}Ga EDTA is used to monitor the impact of DNA associated ^{67}Ga decay.

The results show a consistent difference in relation to analogous experiments with ^{125}I. The amounts of ^{67}Ga activity, as DNA ligand, required to induce ds DNA breaks is substantially more than that for ^{125}I-DNA ligands. Consequently, the controls with the equivalent amount of activity as ^{67}Ga EDTA, show appreciable ds DNA breakage. It is not yet clear whether this reflects a lower DNA binding constant for the ^{67}Ga ligands, or intrinsic properties of the isotopes, or both. However the calculations by Charlton and Humm (personal communication) indicate that the total energy output in the electron cascade is somewhat lower for ^{67}Ga than for ^{125}I, which translates to less frequent outcomes of ds DNA breaks per decay. Nevertheless it is

Figure 5. Induction of ds DNA breaks by a ^{67}Ga-labelled DNA ligands. A desferrioxamine-Hoechst 38317 conjugate was prepared via the isothiocyanate derivative of the bisbenzimidazole and 2 nmoles of it chelated with ^{67}Ga. The resulting chelate containing 0.96 mCi of ^{67}Ga was incubated with 5 µg of pBR322 DNA in 100 µl of a buffer containing 10% methanol in pH 7.4 20 mM citrate for 7 days at 5^{0}C. Controls contained the equivalent amount of ^{67}Ga citrate or DNA and buffer alone (C). Portions (20%) of each incubation mixture were analysed on EtBr-agarose gels (17%) with references of linear (EcoRI cut) pBR322 (R1) and un-incubated pBR322 DNA (R2). The arrows indicate the positions of the origin, nicked, linear and intact pBR322, from top to bottom, respectively.

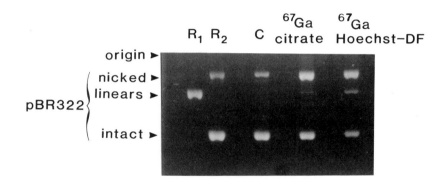

clear from Figure 5 that the ^{67}Ga- DNA ligand shows a much higher level of ds DNA break induction than the equivalent amount of ^{67}Ga EDTA.

DNA sequencing experiments with ^{67}Ga-desferriox-amine-bisbenzimidazole and cytotoxicity experiments with ^{67}Ga-desferrioxamine acridine are underway.

Induction of DNA ds breaks following thermal neutron capture by DNA-bound ^{157}Gd atoms

Boron Neutron Capture Therapy (BNCT) exploits the high thermal neutron capture cross-section of the ^{10}B (n,α) ^{7}Li reaction (Fairchild and Bond, 1985); its effectiveness depends on preferential concentration of ^{10}B in tumour cells where neutron capture generates high LET particles (α and ^{7}Li) with ranges restricted to less than a cell diameter. Clinical trials of BNCT in Japan have shown encouraging results for brain tumours (Hatanaka, 1986) but the potential of the therapy is limited by difficulties in attaining the required levels of ^{10}B in the tumour cells (30 ppm) and poor penetration of the thermal neutron beam.

The rationale for selecting ^{10}B for NCT considers other isotopes with high neutron capture cross-sections, but most of these undergo n,γ reactions and thus don't have the potential of confining the radiation damage so effectively as does the ^{10}B (n,α) reaction. ^{157}Gd has the highest thermal neutron capture cross-section of all naturally occurring isotopes (namely 242,000 barns compared to 3,840 barns for ^{10}B). In view of the role of internal conversion in generating Auger electrons, we decided to explore the possibility that internal conversion may occur to some extent in such n,γ reactions. We were encouraged by the study of Greenwood et al. (1978) which indicated that about 0.88 conversion electrons were emitted per capture event for ^{157}Gd.

Fortunately Gd^{3+} apparently binds to DNA, as it does to nucleotides (Dobson et al., 1978) so our experiments simply involved irradiation of mixtures of Gd^{3+} and pBR322 DNA

Figure 6. Induction of ds DNA breaks by ^{157}Gd neutron capture. Samples of 5 μg of pBR322 in 100 ml of a buffer contained 10 mM each of Hepes and Tris HCl at pH 7.5 (lanes 1-12) or 50 mM Tris HCl (13,14). $GdCl_3$ was added to final concentrations of 0.5 mM (5,9) or 2.5 mM (6,7,8,10,11,12) using either naturally occurring Gd (5,6,7,8) or enriched ^{157}Gd (9,10,11,12). Some samples contained 10 mM EDTA prior to irradiation (2,4,7,11,13,14) and the others had the same amount of EDTA added after irradiation. Unirradiated controls (1,2,8,12) were kept at room temperature while the other samples were irradiated in the neutron column at full reactor power for 5 hrs (13,14) or 6 hrs 40 min (3-7, 9-11).

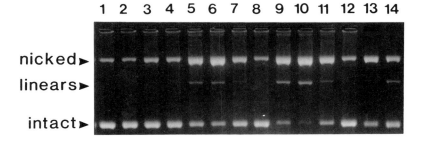

with thermal neutrons and analysis on agarose gels. The results are shown in Figure 6. Control samples were unirradiated (lanes 1, 2, 8 and 12) or irradiated in buffer only (lanes 3, 4 and 13). Lanes 5 and 6 show the effect of inclusion of Gd^{3+} to final concentration of 0.5 mM and 2.5 mM respectively; ds DNA breakage is evident. Lanes 7 and 8 also have 2.5 mM Gd^{3+} but the sample for the former included excess EDTA which reduced ds DNA break induction. Lanes 9-12 are essential repeats of lane 5-8 except using enriched ^{157}Gd (79.7% compared to 15.7%), with consequent increase in DNA damage. We suggest that these results indicate that the ^{157}Gd (n,γ) reaction does produce an electron cascade that induces ds DNA breaks. As in the case of Auger-emitting isotopes, the event needs to take place in close proximity to DNA in order to result in ds DNA break induction; this conclusion is supported by the protective effect of prior addition of EDTA. Lanes 13 and 14 show the result of a separate experiment in which one sample (lane 14) contained borate to a final concentration of 30 mM in the presence of EDTA. In this case the ds DNA breakage is presumably due to the action of high LET particles. The result also highlights the fact that ^{157}Gd is a much more active neutron sensitiser than is boron.

These results have obvious, important implications for neutron radiotherapy.

CONCLUSION

The results discussed here demonstrate the potential in the application of molecular biology techniques to radiobiological questions (Thacker, 1986). In the context of Auger emitters, these methods can contribute to bridging the gap between atomic physics and biological effects.

There is some scope for optimism in the view that DNA ligands labelled with Auger emitters may have a potential role in radiotherapy if combined with suitable delivery systems, such as those that exploit receptor-mediated endocytosis. It is important that it is now clear that such approaches can utilise isotopes other than ^{125}I, in view of the particular problems associated with that isotope. The possibility of generating an Auger cascase by neutron irradiation of a stable isotope, as indicated by our experiments with ^{157}Gd, adds further to the clinical potential.

ACKNOWLEDGEMENTS

The authors thank Drs. D.E. Charlton and J.L. Humm for communicating the results of their calculations on ^{67}Ga decay, and Dr. H. Loewe for the gift of Hoechst 38317.

This work was supported by grants from the Australian Research Grants Scheme and the Australian Institute of Nuclear Science and Engineering.

REFERENCES

Allen, B.J., Brown, J.K., Harrington, B., Izard, B., Linklater, H., Maddelena, D.J., McNeil, J., MacGregor, B.J., Mountford, M.H., Snowdon, G.N., Wilson, D.J., Wilson, J.G., Parsons, P., Moore, D., Tamat, S. & Hersheg, P. In Neutron Capture Therapy, edited by H. Hatanaka (Nishimura), p.258.

Auger, P., 1925, Journal de Physique, 6S, 205.

Auger, P., 1975, Surface Science, 48, 1.

Bloomer, W.D. and Adelstein, S.J., 1977, Nature, 265, 620.

Bloomer, W.D., McLaughlin, W.H., Weichselbaum, R.R., Tonnesen, E.L., Kellman, S., Seitz, D.E., Hanson, R.N., Adelstein, S.J., Rosner, A.L., Burstein, N.A., Nove, J.J. & Little, J.B., 1980, International Journal of Radiation Biology, 38, 197.

Charlton, D.E., 1986, Radiation Research, 107, 163.

Charlton, D.E. and Booz, J., 1981, Radiation Research, 87, 10.

Dervan, P.B., 1986, Science, 232, 464.

Deutzmann, R. and Stöcklin, G., 1981, Radiation Research, 87, 24.

Dobson, C.M., Geraldes, C.F.G.C., Ratcliffe, E. & Williams, R.J.P., 1978, European Journal of Biochemistry, 88, 259.

Fairchild, R.G. and Bond, V.P., 1985, International Journal of Radiation Oncology, Biology and Physics, 11, 831.

Gibbs, R.A., Camakaris, J., Hodgson, E.S. & Maiden, R.F., 1987, International Journal of Radiation Biology, 51, 193.

Greenwood, R.C., Reich, C.W., Baader, H.A., Koch, H.R., Breitig, D., Schultz, O.W.B., Sogelberg, B., Backlin, A., Mampe, W., von Egidy, T., Schreckerbach, K., 1978, Nuclear Physics, A304, 327.

Hatanaka, H., 1986, in Boron Neutron Capture Therapy of Tumours, edited by H. Hatanaka (Nishimura).

Hofer, K.G., Keogh, G. and Smith, J.M., 1977, Current Topics in Radiation Research Quarterly, 12, 335.

Krisch, R.E., 1972, International Journal of Radiation Biology, 21, 167.

Krisch, R.E., Krasin, F. and Sauri, C.J., 1976, International Journal of Radiation Biology, 29, 37.

Krisch, R.E. and Sauri, C.J., 1975, International Journal of Radiation Biology, 27, 553.

Linz, U. and Stöcklin, A., 1986, Radiation Research, 101, 262.

Martin, R.F., 1977, International Journal of Radiation Biology, 32, 491.
Martin, R.F., Bradley, T.R. and Hodgson, G.S., 1979, Cancer Research, 39, 3244.
Martin, R.F. and Haseltine, W.A., 1981, Science, 213, 896.
Martin, R.F. and Holmes, N., 1983, Nature, 302, 452.
Murray, V. and Martin, R.F., 1985a, Gene Analysis Techniques, 2, 95.
Murray, V. and Martin, R.F., 1985b, Nucleic Acids Research, 13, 1467.
Murray, V. and Martin, R.F., 1985c, Journal of Biological Chemistry, 260, 10389.
Schmidt, J., 1984, Journal of Biological Chemistry, 259, 14033.
Schmidt, A. and Hotz, G., 19, International Journal of Radiation Biology, 24, 307.
von Sonntag, C., Hagen, U., Schön-Bopp, A. & Schulte-Frohlinde, D., 1981, Advances in Radiation Biology, 9, 109.
Sundell-Bergman, S. and Johanson, K.J., 1982, Biochemical and Biophysical Research Communications, 106, 546.
Thacker, J., 1986, International Journal of Radiation Biology, 50, 1.

CELL KILLING AND DNA DOUBLE-STRAND BREAKAGE BY DNA-ASSOCIATED ^{125}I-DECAY OR X-IRRADIATION: IMPLICATIONS FOR RADIATION ACTION MODELS

Ian R. Radford and George S. Hodgson

Molecular Science Group
Peter MacCallum Cancer Institute
481 Little Lonsdale Street
Melbourne Victoria 3000
Australia

ABSTRACT

The responses of diploid, tetraploid, and near-hexaploid V79 cells to X-irradiation or DNA-associated ^{125}I-decay were compared. It is concluded that current models of radiation action are unable to explain the observed responses and chromosomal aberration evidence consistent with a new model is presented.

INTRODUCTION

The ability to accurately define the dosage delivered to a cell is an important advantage for researchers using ionizing radiation as a cytotoxic insult. This advantage is magnified by the use of DNA-associated ^{125}I-decay, which allows the delivery of precisely quantifiable levels of damage to the cellular target for radiation action.

Our initial interest was to use DNA-associated ^{125}I-decay as a dosimetric tool for the calibration of a DNA double-strand breakage (dsb) assay (neutral filter elution) (i.e. to determine the relationship between assay measurement and DNA dsb per cell). The basis of this application is the observation that each decay of ^{125}I, incorporated into DNA as iododeoxyuridine (IdU), produces one DNA dsb (Krisch et al., 1977). Initial calibration studies, using mouse L cells and hamster V79 cells, demonstrated a factor of two difference between the assay response of these cells which reflected their relative DNA contents (Radford and Hodgson, 1985). That is, L cells (pseudo-tetraploid) required twice as many ^{125}I-decay-induced DNA dsb as V79 cells (pseudo-diploid) to produce the same level of assay response. This result suggested that the neutral filter elution assay measurement was linearly related to the level of DNA dsb per unit length of

DNA. Because of the importance of this conclusion to further work with the elution assay, and because X-ray studies using both L and V79 cells had shown a similar relationship between the level of elution and the level of cell killing, (Radford and Hodgson, 1985), a study using cells of different ploidy, but derived from the same cell line (V79), was undertaken. The results of this study have been presented elsewhere (Radford and Hodgson, 1987), and are reviewed in this article. The data from the study of ploidy effects suggested a new model of radiation action (Radford et al., 1987) which predicted that the relationship between chromosome aberrations per cell and the level of cell killing should not be affected by ploidy. The results of experiments designed to test this prediction are reported herein for the first time.

MATERIALS AND METHODS
Cells and Neutral Filter Elution

Chinese hamster V79 cells of differing ploidy were produced by treatment with the microtubule-disrupting drug nocodazole. The ploidy of the lines used in this study is defined only on the basis of chromosome number and DNA content. Details regarding cell growth conditions, radioactive labelling, and survival estimation by cloning, have been described elsewhere (Radford and Hodgson, 1985, 1987). ^{125}I-studies were performed by incubating cells with different concentrations of ^{125}IdU for just over one population doubling time, then determining their specific activity, followed by freezing and storage over liquid nitrogen to accumulate a known number of radioactive decays prior to elution and survival measurements (Radford and Hodgson, 1985, 1987).

Elutions were performed as outlined previously (Radford and Hodgson, 1985). Mouse L cells, labelled with tritiated thymidine [^{3}H]dT and given 45 Gy of X-irradiation, were used as an internal reference and were mixed with each test sample of [^{14}C]Thd- or ^{125}IdU-labelled cells prior to lysis and elution. The results are then expressed as relative elution, which is the ratio of the fractions of the total ^{14}C or ^{125}I c.p.m. to ^{3}H c.p.m. eluted from the filter after 17 hr of pumping.

Chromosome aberration studies

In order to avoid the complications which arise from the use of asynchronous populations, these studies were performed with synchronized cells. Mitotic diploid or tetraploid V79 cells, collected following treatment of asynchronous cultures with nocodazole, were rinsed, and then placed in fresh growth medium and incubated at 37^0C for 1 hr. As shown previously (Radford and Broadhurst, 1986), this protocol produces cell populations that are synchronized in G_1/early S-phase. These cells were placed on ice for 30 min, X-irradiated, and then returned to 37^0C. Survival was measured on replicate cultures immediately after irradiation. Cells in the first mitotic wave were collected for chromosome aberration scoring by the addition of nocodazole to 0.1 µg/ml for 3 hr at 12-15 (diploid) or 14-17 (tetraploid) hr after irradiation. Cells were processed for chromosome aberration scoring according to the method of Peterson et al., (1979).

RESULTS AND DISCUSSION
X-ray response and ploidy

The survival response of diploid, tetraploid, and near-hexaploid V79 cells to X-irradiation is shown in Figure 1. The hexaploid cells are considerably more sensitive than either the diploid or tetraploid lines which both show similar responses. DNA dsb induction by X-rays was assayed on replicate cultures to those used for the survival studies (Figure 2). The results mirror the survival data, in showing that the hexaploid cells are more sensitive to DNA dsb induction than the diploid or tetraploid. We have no explanation for this difference, other than to suggest that levels of critical sulphydryl-containing radioprotective compounds may be lower in the hexaploid line.

When the level of X-ray-induced lethal lesions (-ln survival) was plotted against DNA dsb (relative elution), the relationships for each cell line were linear and not significantly different from each other (Figure 3). This is consistent with previous data which suggested a general relationship between X-ray-induced cell killing and DNA dsb for mammalian cells (Radford, 1986). Based on the observation that relative elution reflects the level of DNA dsb per unit length of DNA (Radford and Hodgson, 1985), the data also suggest a direct relationship between DNA dsb/lethal lesion and ploidy.

Effect of ploidy on ^{125}I-decay response

Similar studies were conducted using DNA-associated ^{125}I-decays to induce cell killing and DNA dsb. The survival response data are shown in Figure 4, and, consistent with the X-ray data, they show that tetraploid cells (D_0 = 121 $\pm$ 4 decays) require twice as many ^{125}I-decays (which can be equated with DNA dsb) to produce a lethal event as do diploid cells (D_0 = 60 $\pm$ 1 decays).

Figure 1. Clonogenic survival of diploid (▲), tetraploid (■), and hexaploid (●) V79 cells as a function of X-ray dose.

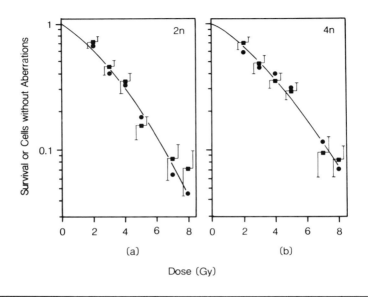

Figure 2. DNA dsb induction dose-response curves for X-irradiated diploid (▲), tetraploid (■), and hexaploid (●), V79 cells, as assayed by neutral filter elution.

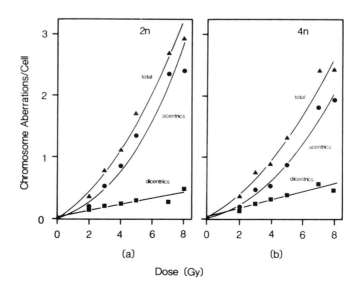

However, the hexaploid cells (D_0 = 137 $\pm$ 5 decays) deviated from this relationship. We cannot explain the hexaploid cell result, except to suggest that they are more prone to misrepair DNA dsb.

The effect of ploidy on the DNA dsb assay response, following ^{125}I-decay accumulation in cells, is shown in Figure 5. As expected on the basis of previous data (Radford and Hodgson, 1985), the sensitivity of diploid cells was 1.96 $\pm$ 0.05 and 2.95 $\pm$ 0.10 times that of tetraploid and hexaploid cells respectively; which is in agreement with the relative DNA contents of these lines of 1.98 $\pm$ 0.07 and 2.79 $\pm$ 0.10. This result confirms the conclusion that relative elution is linearly related to the level of DNA dsb per unit length of DNA.

When the level of ^{125}I-decay induced lethal lesions was plotted against relative elution (Figure 3), the relationship for each of the ^{125}I-decay data sets was significantly different to that for the corresponding X-ray data set. This suggests that an ^{125}I-induced DNA dsb is more likely to produce a lethal event than is an X-ray-induced DNA dsb. Assuming that both types of DNA dsb are randomly distributed through the genome, one possible explanation for this difference in lethal efficiency, is that there is a discrepancy between the cellular repair systems and the DNA dsb assay technique as to what lesions are recognized as "DNA dsb". For example, the neutral filter elution assay system does not discriminate between restriction endonuclease-induced blunt-ended and cohesive-ended DNA dsb (Radford, unpublished data), although cellular repair systems can differentiate between them (Bryant, 1984).

Implications for radiation action models

We are unable to reconcile the above results with the suggestion that ionizing radiation-induced cell death is due to the unmasking of recessive lethal mutations or gene dosage effects caused by chromosome fragment loss (e.g. Hopwood and Tolmach, 1979). The finding that the number of DNA dsb per cell required to produce a lethal lesion is proportional to ploidy makes the loss of the same region of the corresponding homologues from each chromosome set in polyploid cells appear extremely unlikely. To explain these (and other) results we have postulated a new model of radiation action termed the "Critical DNA target size" model (Radford et al., 1987). This model suggests that DNA dsb is the critical radiation-induced lesion and that DNA dsb within critical regions (targets) of the nuclear DNA produces lethal lesions. Mathematical analysis showed that

Figure 3. The relationship between the levels of neutral filter elution and lethal lesions (-ln survival) for X-rays (solid symbols) and ^{125}I-decay (open symbols): diploid (▲), tetraploid (■), and hexaploid (●) V79 cells. Linear regression estimates of the line of best fit for each data set were determined. Analysis of covariance showed no significant difference between the slopes calculated for each X-ray data set (P = 0.22). The slope values were: 2n = 35.36 ± 1.64 (s.e.m.); 4n = 38.74 ± 1.28; and 6n = 37.61 ± 1.34. However, there were significant differences between the respective X-ray and ^{125}I-decay data sets (P<0.05). The slope values for the ^{125}I-decay data sets were: 2n = 54.38 ± 5.35; 4n = 49.26 ± 2.96; and 6n = 67.74 ± 8.02.

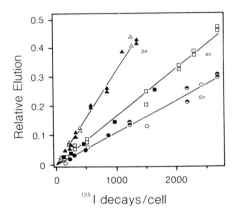

Figure 4. Clonogenic survival of V79 lines of differing ploidy following DNA-associated ^{125}I-decay accumulation during storage over liquid nitrogen: diploid (△ , ▲), tetraploid (□ , ■), and hexaploid (○ , ◐ , ●).

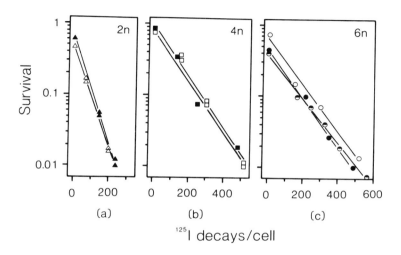

Figure 5. The effect of ploidy on the neutral filter elution assay response of cells containing ^{125}I-induced DNA dsb.

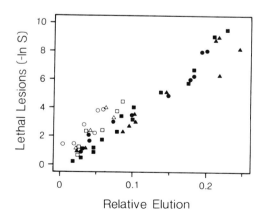

the latter suggestion could be reconciled with the polyploid cell results if the target size was independent of ploidy. The probability of conversion of a DNA dsb, occurring in a target, into a lethal event should then be independent of ploidy. Assuming that chromosomal aberrations are directly related to cell killing, our model predicts that the relationship between aberrations per cell and level of killing should not be affected by ploidy.

Survival and chromosome aberrations in cells of differing ploidy

To test the above prediction we compared the dose-response relationships for X-ray-induced chromosome damage in cells of different ploidy. Consistent with most cells being synchronized in G_1/early S-phase at the time of X-irradiation, very few chromatid-type aberrations were observed. The dose-response relationships for acentrics, dicentrics, and total chromosome-type aberrations per diploid or tetraploid cell are shown in Figure 6. It is apparent that the acentric fragment is the predominant aberration induced by X-rays in these cells. The data for both aberration types, fitted to a quadratic function of dose, show qualitatively similar responses for diploid and tetraploid cells (see legend of Figure 6). Acentric fragment induction is greater in diploid than in tetraploid cells, consistent with the relatively higher level of cell killing found for replicate diploid cultures (Figure 7).

Clonogenic survival and the fraction of cells without chromosome-type aberrations, were plotted against X-ray dose (Figure 7). For both diploid and tetraploid V79 cells, there is a close agreement between these two measurements. Similarly, the fraction of cells predicted to lack chromosome aberrations, assuming total chromosome-type aberrations have a Poisson distribution, was consistent with the above-mentioned measurements (data not shown but can be derived from Figure 6).

We are not aware of any similar comparisons of diploid and tetraploid cells in the literature. Greenblatt (1962) and Roberts and Holt (1982) have examined chromosomal aberration induction by low LET radiation in diploid V79 cells. Our aberration dose-response curves are very similar to those of Greenblatt (1962), but differ from those of Roberts and Holt (1982) in that they found similar dose-response curves for the production of acentric fragments and dicentrics by ^{60}Co gamma irradiation. This difference might relate to the use of cells synchronized by medium depletion in the latter study. A close agreement

Figure 6. Dose-response relationships for the induction of acentric fragments (●), dicentrics (■), and total chromosome-type aberrations (▲) (i.e. acentric fragments, dicentrics, and centric rings) in X-irradiated diploid (a) and tetraploid (b) V79 cells. At least 100 cells were scored at each dose. The curves for the acentric fragment and total aberration data are computer-generated fittings to an equation of the form $Y = a + bD + cD^2$, where Y is chromosomal aberrations per cell, D is dose, and a, b, and c are constants. The dicentric fragment data were fitted to the equation $Y = a + bD$ (by regression analysis), as negative values for c were obtained when the quadratic equation was used. The calculated values ($\pm$ s.e.m.), of the constants a, b, and c (or a and b), for diploid cells were: acentrics (1.59 $\pm$ 0.15 x 10^{-2}, 3.96 $\pm$ 0.49 x 10^{-2}, and 3.84 $\pm$ 0.14 x 10^{-2}), dicentrics (4.03 x 10^{-2}, and 4.88 $\pm$ 0.90 x 10^{-2}), and total (1.99 $\pm$ 1.27 x 10^{-2}, 1.43 $\pm$ 0.34 x 10^{-1}, and 3.18 $\pm$ 0.57 x 10^{-2}). For tetraploid cells the calculated values were: acentrics (2.45 $\pm$ 0.44 x 10^{-2}, 3.97 $\pm$ 0.72 x 10^{-2}, and 2.62 $\pm$ 0.13 x 10^{-2}), dicentrics (4.37 x 10^{-2}, and 6.56 $\pm$ 0.85 x 10^{-2}) and total (4.29 $\pm$ 0.37 x 10^{-2}, 1.30 $\pm$ 0.04 x 10^{-1}, and 2.34 $\pm$ 0.09 x 10^{-2}).

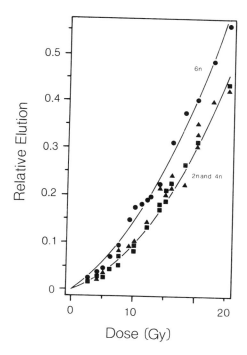

Figure 7. Dose-response relationships for clonogenic survival (●) and fraction of cells without asymmetrical chromosome aberrations (■) in X-irradiated diploid (a) and tetraploid (b) V79 cells. The curves shown are fittings to the survival data. The fraction of cells without aberrations is expressed relative to the control value. The uncertainty bars are one standard error. For the survival data, the standard error of the mean was not appreciably larger than the plotted symbol.

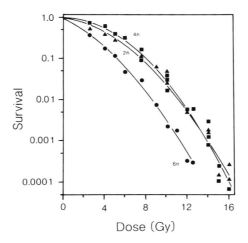

between the fraction of diploid or tetraploid cells lacking chromosomal aberrations and clonogenic survival was found. This finding has been reported by a number of groups using diploid CHO or V79 Chinese hamster cells (Roberts and Holt 1982, reviewed in Chadwick and Leenhouts, 1981) and Syrian hamster fibroblasts (Roberts and Holt, 1985); however, in the context of an overall explanation for the results of this study, we suggest that it is highly significant that tetraploid cells also show this agreement.

The results of these preliminary chromosomal aberration studies are thus consistent with one of the predictions of the "Critical DNA target size" model (Radford et al. 1987).

REFERENCES

Bryant, P.E., 1984, International Journal of Radiation Biology, 46, 57.

Chadwick, K.H., and Leenhouts, H.P., 1981, The Molecular Theory of Radiation Biology (Berlin: Springer-Verlag), pp.120-126.

Greenblatt, C.L., 1962, International Journal of Radiation Biology, 4, 185.

Hopwood, L.E., and Tolmach, L.J., 1979, Advances in Radiation Biology, 8, 317.

Krisch, R.E., Krasin, F., and Sauri, C.J., 1977, Current Topics in Radiation Research Quarterly, 12, 355.

Peterson, W.D., Simpson, W.F., and Hukku, B., 1979, Methods in Enzymology, 58, 164.

Radford, I.R., 1986, International Journal of Radiation Biology, 49, 611.

Radford, I.R., and Hodgson, G.S., 1985, International Journal of Radiation Biology, 48, 555.

Radford, I.R., and Broadhurst, S., 1986, International Journal of Radiation Biology, 49, 909.

Radford, I.R., and Hodgson, G.S., 1987, International Journal of Radiation Biology, 51, 765.

Radford, I.R., Hodgson, G.S., and Matthews, J.P., 1987, International Journal of Radiation Biology, manuscript submitted.

Roberts, C.J., and Holt, P.D., 1982, International Journal of Radiation Biology, 41, 645.

Roberts, C.J., and Holt, P.D., 1985, International Journal of Radiation Biology, 48, 927.

NEW CONCEPTS OF DNA FUNCTIONING AS REVEALED BY THE LETHAL EFFECT OF ^{64}Cu AND ^{67}Cu DECAYS

S. Apelgot* and E. Guille**

*Institut Curie, Section de Physique et Chimie 11 rue P. et M. Curie, F-75231 Paris Cedex 05

**Université Paris-Sud, Biologie Moléculaire
Végétale (UA 1128) F-91405 Orsay Cedex

ABSTRACT

It is well known that under the same experimental conditions, the lethal effect of decay of 2 different radioisotopes of iodine, ^{125}I and ^{131}I, is very different only because of their different decay schemes (β, γ for ^{131}I and electron capture for ^{125}I). To pursue these experiments, we tested radioactive isotopes of copper, ^{64}Cu and ^{67}Cu. ^{67}Cu has a decay scheme comparable to that of ^{131}I while 50% of ^{64}Cu decay via electron capture. In spite of their different decay schemes, both copper isotopes provoked the same lethal effect under the same experimental conditions. The survival curves are exponential for mammalian cells and the lethal efficiency is high. This result demonstrates that the decay inside the DNA molecule of only a few radioactive copper atoms, regardless of their decay schemes, has a consequence as severe as mammalian cell death. To understand this surprising result, we suggest a microcomputer-like model for the organization and functioning of molecular DNA knowing that metals at trace levels are one of the key-points of microcomputers. The three major components (control, processing and exchange units) of a classical microcomputer were defined. This model is able to shed new light on the non-uniform radiosensitivity of mammalian cellular DNA to ^{125}I decays utilized in the form of ^{125}IdU.

INTRODUCTION

As early as 1970, Ertl et al. suggested that the lethal efficiency of the decay of ^{125}I incorporated into cellular DNA should be very high because of the particular decay scheme of this radioactive isotope. This hypothesis was widely confirmed by Hofer & Hughes, (1971); Burki et al., (1973); Chan et al., (1976); Warters et al., (1977) and Ritter, (1980, 1981). ^{125}I disintegrates by electron capture and internal conversion, resulting in a multiionized ^{125}Te atom, the positive charge of which is equivalent to about 300 eV and in the emission of about 20 Auger electrons of low energy from 0.8 to 34.6 eV (Charlton et al., 1978; Charlton, 1986).

When ^{125}I is incorporated in synchronized mammalian cells, the survival curves are exponential, and the lethal efficiencies are high only when ^{125}I decays have occurred in the DNA molecule itself (Warters et al., 1977; Commerfold et al., 1980). Moreover, this high lethal efficiency is, in fact, a consequence of the particular decay scheme of ^{125}I. I^{131} which disintegrates by emission of β-particles only (Figure 1), gives survival curves with a large shoulder when incorporated like ^{125}I, in the form of ^{131}I-dU in cellular DNA. Therefore the lethal efficiency of its decay is much lower than that of ^{125}I (Hofer & Hughes, 1971; Bradley et al., 1975; Chan et al., 1976). It must be emphasized that when used as the same molecule and under identical experimental conditions, the decays of ^{125}I and ^{131}I provide very different results with regard to the lethal effect only because their disintegration schemes are different.

To test whether these differences between two different radioisotopes of iodine, due to their different disintegration schemes, are a general phenomenon, two radioactive isotopes of another element, copper, were studied. This element was chosen because it is present in trace amounts in the DNA of mammalian cells (Guillé et al. 1981), and since two isotopes of copper, ^{64}Cu and ^{67}Cu, differing in their disintegration scheme, exist. 41% of ^{64}Cu decays by means of electron capture and 87% of ^{67}Cu via β-emission (Figure 1). The differences in the decay schemes of these two copper radioisotopes are fairly comparable to those of ^{125}I and ^{131}I (Figure 1). Surprisingly, the lethal effect of the decay of ^{64}Cu or ^{67}Cu atoms was the same under the same experimental conditions. To understand this unexpected result, a microcomputer-like model for the organization and functioning of molecular DNA was suggested. This model is based on the fact that metals at trace levels are one of the key-points of microcomputers, as they appear to be for the DNA functioning.

REPORT ON PREVIOUS RESULTS

Experiments (Apelgot et al., 1984, 1987) were performed

with two types of mammalian cells, A549, a human malignant cell line, and CV1, a simian cell line. ^{64}Cu and ^{67}Cu were used as chloride salt. Unlabelled $CuCl_2$ was always added in order to obtain a standard final copper concentration of 1 to 2 µg/ml. These experiments have shown that the decay of the two radioactive isotopes of copper, ^{64}Cu and ^{67}Cu, incorporated in non-synchronized mammalian cells leads to a lethal effect characterized by an exponential survival curve (Apelgot et al., 1984, 1987). The salient result was that these two radionuclides have identical and high lethal efficiencies despite their different disintegration schemes. It was also observed that the lethal event occurs inside the DNA molecule of the mammalian cells studied and that this lethal event is irreparable (Apelgot et al., 1987). These unexpected results led us to conclude that the lethal effect of ^{64}Cu or ^{67}Cu decays is a consequence of the decay phenomenon itself. For all radioactive atoms, the decay phenomenon takes place over a very short time (10^{-15} sec.) and corresponds to an abrupt and sudden modification of an unstable atom nucleus. These results focus on the fact that the decay of only a few radioactive copper atoms has a consequence as severe as the death of a mammalian cell. Recently, we have also observed a lethal effect as a consequence of the decay of radioactive zinc atoms (^{65}Zn) incorporated in mammalian cells (Joseph, 1984). The question is thus as follows: what might the fundamental role of copper and zinc inside the DNA molecule be, since a local disturbance in these metal atoms brings about the death of mammalian cells with a high efficiency? Given this cause-effect relationship (radioactive metal decay/cell death), one might think that a fundamental structure containing copper and/or zinc is implicated in the management of the overall behaviour of cells. Copper and zinc have natural access to this structure, probably in the chromatin areas of mammalian cells. Their concentration in this structure depends on cellular type (Tashima et al., 1981), as well as on environmental variations (Guille et al., 1981).

It is well known that metals, at trace levels, are one of the fundamental parts of microcomputers. We would like to suggest, as a working hypothesis, that metals, also at trace levels, might be, as in a microcomputer, one of the key-points for cellular DNA functioning.

THE SUGGESTED MODEL

Based on the microcomputer scheme (Apelgot et al., 1987) we tried to foresee a cell organization able to fulfill the basic components of a microcomputer (Figure 2). Each

Figure 1. Decay schemes of ^{64}Cu, ^{67}Cu, ^{125}I and ^{131}I. E.C. = electron capture; I.C. = internal conversion. (From Lederer and Shirley 1978).

$^{64}_{28}$Ni $\longleftarrow$ $^{64}_{29}$Cu $\longrightarrow$ $^{64}_{30}$Zn β^- : 0.573 MeV
(12.8h)

19% β^+ 40% β^- β^+ : 0.659 MeV
41% E.C.

$^{67}_{29}$Cu $\longrightarrow$ $^{67}_{30}$Zn + β^- + γ β^-
62h

1% : 0.190 MeV
56% : 0.400 MeV
23% : 0.480 MeV
20% : 0.580 MeV

6 γ from 0.09 to 0.4 keV

$^{131}_{53}$I $\longrightarrow$ $^{131}_{54}$Xe + β^- + γ β^-
8.04d

87.2% : 0.608 MeV
9.3% : 0.320 MeV
2.8% : 0.250 MeV
0.7% : 0.810 MeV

4 γ from 0.08 to 0.638 MeV

$^{125}_{52}$Te $\longleftarrow$ $^{125m}_{52}$Te $\longleftarrow$ $^{125}_{53}$I

I.C. 93%   E.C.   7% γ of 0.035 MeV
γ 7%
1.6 ns   60d

signal (intra or extracellular) is received by the corresponding receptor (exchange unit : membranes, DNA). These receptors transform the signal (copy and translation) in such a way that via this transformation the message can then be received by the control unit. In DNA molecules, the role of the 'control unit' may be played by the highly reiterative DNA sequences preferentially localized at the periphery of the nucleus (Williams & Ockey, 1970). The

---

Figure 2. Microcomputer-like functioning of a cell. Distribution of cellular DNA among three major compartments: the exchange-like unit, the processing-like unit and the control-like unit. (A): translation and copy of the initial message; this step includes translation and transformation. (B): file interaction: loading and storing. (C): control and mangement. Satellite DNA is a DNA component which sediments as a satellite band of the main DNA band in CsCl or Cs_2SO_4 density gradients.

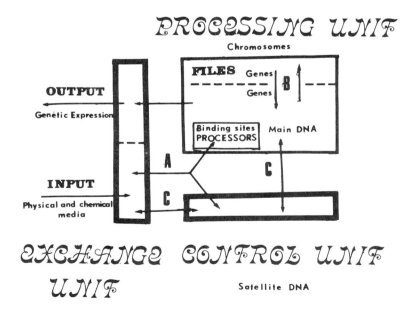

message is then transmitted to the 'processing unit' which is the structural support of genetic material. The cell can thus perform the requested program which is the answer to the initial signal. Such an answer (reorganization of DNA sequences and/or protein synthesis) is reached at the 'exchange unit' level (ribosome, synthesis cycle).

The various inter-relations of this model given in Figure 2 are a good illustration of a general concept of membrane impression leading to genetic expression first proposed some years ago (Witham et al., 1978).

CONSEQUENCES OF OUR MODEL

Our model implies that the same kind of lesions in the different types of DNA must have different consequences with regard to the DNA role in the 'processing unit'. If the lesion occurs inside the genome (single copy DNA sequences), the capacity for information processing will only be decreased. Conversely, if the lesion occurs inside the control unit (highly reiterative DNA sequences), the processing unit will be completely disturbed, either by a signal interruption, or by an incorrect signal. This model explains the variations evidenced by Burki et al. (1977) in the lethal efficiency of ^{125}I decay when ^{125}I-deoxyuridine was used at different steps of the S phase of synchronous Chinese hamster cells. For cells labelled in the DNA replicated in the first quarter of the DNA cycle (synthesis of unique and moderately reiterative DNA sequences), 62 ^{125}I decays are needed to kill a cell, while only 16 decays are needed when DNA is replicated in the second half of the DNA synthesis period (synthesis of satellite DNA : control unit). These results were recently confirmed by Yasui et al. (1985); they have also shown that lethality of the decays of ^{125}I incorporated in the form of ^{125}IdU in the DNA is a consequence of ^{125}I atoms located only in a small part of the DNA and not of any ^{125}I decay.

In the case of ^{64}Cu and ^{67}Cu, it was observed that for mammalian cells under in vitro growth conditions the lethal effect is correlated to these radioactive atoms entering inside the DNA molecule at a late time after the beginning of the contact between the cells and the isotopes, that is 22 hours and 26 hours for A549 and CV1 cells respectively (Apelgot et al., 1987). These late times are in agreement with the critical period evidenced at the end of the S phase for ^{125}IdU (Burki et al., 1977).

All of these results suggested that, for mammalian cells under in vitro growth conditions, radioactive copper atoms and ^{125}IdU molecules have to reach particular sites and these sites only, inside the control unit, in order to give rise to lethal events via decays. Therefore the 'death target' suggested by Yasui et al., (1985) would clearly be located inside the control unit.

CONCLUSIONS

In order to understand why the lethal effect of two different radioactive isotopes of copper is the same in spite of their different decay schemes, we suggest a microcomputer-like model for the organization and functioning of molecular DNA. This model allows an understanding of why the lethal effect of the decays of ^{125}I utilized in the form of ^{125}IdU is related to the decays occurring inside particular parts of the DNA of mammalian cells under in vitro growth conditions; such specific sites are replicated in the second half of the DNA synthesis period. According to our model, this particular DNA would correspond to the most important part of the microcomputer: the control unit.

REFERENCES

Apelgot, S., Coppey, J., Grisvard, J., Guillé, E. & Sissoëff, I., 1984, Comptes Rendus de l'Académie des Sciences, Paris, 298 III, 31.

Apelgot, S., Guillé, E., and Cols, P., 1987, (submitted to Progress in Molecular Biology).

Apelgot, S., Coppey, J., Gaudemer, A., Grisvard, J., Guillé, E., Sasaki, I., and Sissoëff, I., 1987, (submitted to Journal of Cell Science).

Bradley, E.V., Chan, P.C. & Adelstein, S.J., 1975, Radiation Research, 64, 555.

Burki, H.J., Roots, R., Feinendegen, L.E. & Bond, V.P., 1973, International Journal of Radiation Biology, 24, 363.

Burki, H.J., Koch, C. & Wolff, S., 1977, Current Topics in Radiation Research, Vol. 12, edited by M. Ebert and A. Howard (North-Holland, Amsterdam), p.408.

Chan, P.C., Lisco, E., Lisco, H. & Adelstein, S.J., 1976, Radiation Research, 67, 332.

Charlton, D.E., Booz, J., Fidorra, J., Smit, Th. & Feinendegen, L.E., 1978, Proceedings of the 6th Symposium on Microdosimetry, edited by J. Booz and H.G. Ebert (Harwood Academic Publishers Ltd) p.91.

Charlton, D.E., 1986, Radiation Research, 107, 163.

Commerfold, S.L., Bond, V.P., Cronkite, E.P. & Reincke, U., 1980, International Journal of Radiation Biology, 37, 547.

Ertl, H.H., Feinendegen, L.E. & Heiniger, H.J., 1970, Physical and Medical Biology, 15, 447.

Guillé, E., Grisvard, J. & Sissoëff, I., 1981, Systemic Aspects of biocompatibility, Vol. 1, edited by D.F. Williams (CRC Press, Palm Beach, USA) p.39.

Hofer, K.G. & Hughes, W.L., 1971, Radiation Research, 47, 94.

Joseph, A., 1984, Thesis (Paris).

Lederer, C.M. & Shirley, V., 1978, Table of Isotopes, 7th edn (J. Wiley & Son).

Ritter, M.A., 1980, Radiation Research, 84, 113.

Ritter, M.A., 1981, Biophysica Biochemica Acta, 652, 151.

Tashima, M., Calabretta, B., Torelli, G., Scofield, M., Matzel, A. & Saunders, F., 1981, Proceedings of the National Academy of Sciences (USA), 78, 1508.

Warters, R.L., Hofer, K.G., Harris, L.R. & Smith, J.M., 1977, Vol. 12, Current Topics in Radiation Research, edited by M. Ebert and A. Howard (North-Holland, Amsterdam) p.389.

Williams, C.A. & Ockey, C.H., 1970, Experimental Cell Research, 63, 365.

Witham, F.H., Hendry, L.B. & Chapman, O.L., 1978, Origins of life, 9, 7.

Yasui, L.S., Hofer, K.G. & Warters, R.L., 1985, Radiation Research, 102, 106.

CALCULATION OF SINGLE AND DOUBLE STRAND DNA BREAKAGE FROM INCORPORATED ^{125}I

D.E. Charlton

Concordia University,
1455 de Maisonneuve Blvd.,
Montreal, Quebec,
Canada. H3G 1M8

ABSTRACT

A computer code simulating electron tracks is used to follow the electrons released by individual decays of ^{125}I incorporated into double stranded DNA. The tracks (locations of ionisations and excitations) are superimposed on a three dimensional model of the DNA molecule simulating the bases and the two spiraling sugar-phosphate chains. An empirical choice of 17.5 eV deposited in one sugar-phosphate moiety appears to explain the production of a single break, i.e. good agreement with single strand fragmention measured by Martin and Haseltine (1981) is obtained.

This technique which produces a very complex pattern of energy deposition in the DNA is analysed for the production of breaks of several kinds. A major result is that single or double strand breakage depends not only upon the energy deposited in the DNA but also its distribution. Also it will be shown that energy depositions in the DNA below 50 eV produce little or no effect, and that double strand breaks require at least 100 eV.

INTRODUCTION

The cascade of electrons following the electron capture and internal conversion which is the mode of decay of ^{125}I is well documented (Feinendegen, 1975; Hofer, 1980). The number of electrons comprising the cascade have a wide range of number and energy. Charlton and Booz, (1981) showed that between two or three and up to fifty electrons may be released by the decay, and the energy of these electrons may lie between a few eV and several tens of kiloelectron volts.

A segment of DNA containing the decay of incorporated ^{125}I will then experience very different radiation effects, governed by chance, depending upon which of the many possible cascades occurs. Further, neighbouring DNA not containing the decay will be exposed to the longer ranged K- and L-Auger electrons which, while resembling low LET radiation, are capable of producing strand breaks. There will also be a dose produced by the characteristic x-rays and the gamma-ray liberated by the decay. Thus the simple idea of one double strand break per decay at the site of the decay is an inadequate view of the radiation effect of incorporated ^{125}I. Certainly most of the decays will produce considerable damage to the molecule at the site of the decay, including vapourisation (Linz and Stöcklin, 1985), however others may produce little or no damage there but contribute to a low LET background irradiation of the cell nucleus. These various contributions to the radiation effect are illustrated in the first figure.

The packing of the DNA may also be important. Clearly if the DNA molecule is condensed, as in chromatin, then the probability of the longer ranged electrons producing damage is greater than if the DNA is relaxed and diffused throughout the nuclear medium. Further, the arrangement of the DNA, whether it is bound to nucleosomes or supercoiled about itself, may produce different stresses in the molecule following strand damage. Thus the rapid uncoiling of supercoiled plasmids may produce damage not present with

Figure 1. Illustrating the various contributions to the radiation dose from incorporated ^{125}I.

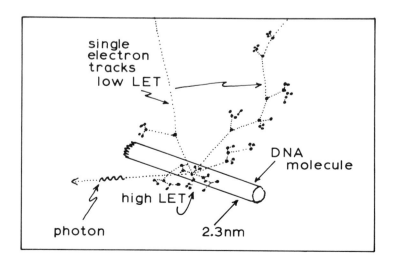

nucleosome-bound DNA.

In previous treatments of the radiation effects of incorporated ^{125}I the average dose to the cell nucleus has been used to evaluate an RBE for the radiation (Kassis et al., 1987; Pomplun, 1987). Typically this is in the order of 8 for cell death. RBE however, while useful for radiation protection or therapy and indicative of differences between radiations, does not provide an insight into the mechanisms of radiation damage. An approach at the molecular level may yield information on this. A microscopic approach is now possible using the technique of track structure codes which simulate the passage of electrons in water interaction by interaction (Paretzke, 1987, for example). Several authors (Chatterjee and Magee, 1985; Ito, 1987) using these codes and models of DNA have examined low LET radiations using single and double strand breaks as a measure of effect. In this presentation, an electron track structure code is used in combination with a simple model of DNA and individual Auger electron spectra to study the damage to the DNA near incorporated ^{125}I. Here the experiment of Martin and Haseltine (1981) provides a valuable test of energy deposition on a molecular scale and it is used to evaluate a mechanism for single strand breakage. By combining single strand breaks on opposite strands, the production of double strand breaks can be predicted. This technique allows the large variability of the decay to be handled, at least at the site of the decay. Remote from the decay, the contribution of the background radiation from the decay can be estimated from published experimental data.

INDIVIDUAL AUGER ELECTRON SPECTRA

Two sources of electron spectra were used in this calculation. These differ in their treatment of the low energy electron component of the spectra in two ways. In the original work (Charlton and Booz, 1981; Humm, 1984) it was assumed that the electron energies were given by the "Z+1" approximation (Coghlin and Clausing, 1973) and that these energies were uneffected by the state of ionisation of the atom. More recently Pomplun et al., (1987) have used a considerably more sophisticated model in which the energy of the electron was evaluated from the difference between the energies of the atom before and after the emission of the Auger (or Coster-Kronig) electron. Generally these electrons are lower in energy than those given by the Z+1 approximation.

Secondly it is likely that the cascade is sufficiently fast so that the daughter tellurium atom charges up due to the emission of electrons and that the subsequent neutralisation of the atom produces further liberation of electrons (Carlson and White, 1963). Charlton and Booz used two extreme cases in that either it was assumed that the iodine atom was isolated (with no chance of neutralisation) or that the valence shell contained a large number of electrons and in effect the atom remained neutral. The average number of electrons released in these two cases was 13 and 21 respectively. The additional number released for the "condensed phase" (large number of electrons in outer shells) were primarily of energy less than 100 eV. These electrons were assumed to be equivalent to those released by the neutralisation processes. In the treatment of Pomplun et al., (1987) only the electrons emitted from the isolated atom are dealt with and it is assumed that the energy present as the charge on the daughter atom (about 1 keV per decay on average) is deposited locally. The model therefore predicts the release of fewer electrons per decay (13).

One thousand individual electron spectra are available from each of these two methods, both of which are used in what follows.

THE ELECTRON TRACK STRUCTURE CODE

The code used here is MOCA7B developed by Paretzke (1987) and is based on cross-sections for electrons passing through unit density water vapour. The code simulates the tracks of electrons, including all of the delta-rays, giving the co-ordinates of each interaction and the energy deposited in the form of excitations and ionisations. To use the code for an incorporated ^{125}I decay the origin of each of the electrons was set to the position of the iodine atom in the DNA molecule, and the initial direction of each electron was chosen to be randomly isotropic. Each decay could be repeated by choosing a new set of initial directions and running the code with a new set of random numbers. The electrons were followed out to 60 nm from their origin. The positions of the energy depositions and the energies were stored for each decay and superposed on the model of DNA described in the next section.

MODEL OF THE DNA

A volume model of the DNA is shown in Figure 2. The molecule is assumed to be a straight solid cylinder of diameter 2.3 nm and divided into 0.34 nm slices. Each slice (shown in the Figure) is divided into three volumes; a central cylinder of diameter 1.0 nm being approximately the volume occupied by a pair of bases, and two arches representing the volumes occupied by the sugar-phosphate molecules. For sequential nucleotide pairs these arches are rotated by 36 degrees simulating the twist of the sugar-phosphate chain. These dimensions can be found in Kornberg (1974) for example. The incorporated iodine atom is placed 0.15 nm from the central axis of the base cylinder symmetrical to one of the arches.

This simple model of the DNA molecule allows a rapid scoring of the interactions of the electrons originating within it. It is assumed that interactions occurring in the grooves of the DNA are as effective at producing damage as those interacting with the atoms of the DNA itself. Further, interactions outside the volume, producing radicals, are ineffective. To what extent these two assumptions compensate for each other is unknown but later (next section) a normalisation process will reduce the importance of this approximation.

COMPARISON WITH THE DATA OF MARTIN AND HASELTINE

The calculations of Chatterjee and Magee, (1985) and Ito, (1987) on strand breakage were based either on direct interaction with the atoms of the DNA alone or combined with OH radical interaction with the DNA backbone. The check of their methods was either the yield of breaks (Chatterjee and Magee) or the RBEs of various low LET radiations (Ito). For incorporated ^{125}I irradiation a much more direct test of strand breakage is possible using the

Figure 2. Model of one nucleotide pair showing the volumes occupied by the bases and the sugar-phosphate molecules.

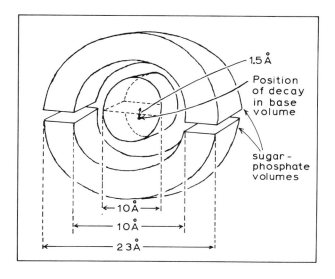

experimental data of Martin and Haseltine. These authors measured the distribution of distances to the single strand breaks most remote from iodine decays. Their results are shown in Figure 3 for breaks in the strand bonded to the based containing the decay. These data will now be used to establish a criterion for single strand breaks.

From the origin of the decay in one of the bases, the electron tracks for individual decays were generated and the energy deposited in the volumes representing the components of the DNA recorded. Table 1 gives the results for three individual decays. In the Table the upper string of figures is the energy depositions in individual sugar-phosphate volumes, one of which is bonded to the base containing the decay. The central line is the energy depositions in the base pairs and the lower is that for the complementary strand. The code for the pixels is given in the table and the position of the decay is indicated in the table by "↑".

Figure 3. The distribution of farthest distances to single strand breaks in the strand bonded to the base containing the decay.

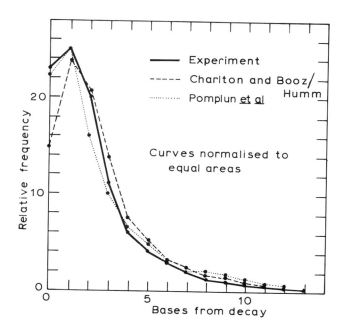

Table 1. Energy deposition by single decays in sugar-phosphate and base volumes.

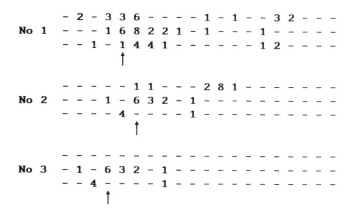

Pixel code: "-" Energy deposition less than 10eV; "1" Energy deposition between 10 and 20 eV; "2" Energy deposition between 20 and 30 eV; etc.

To simulate the Martin/Haseltine experiment assume for the moment that single strand breaks to the right of the decay can be detected in the experiment and that 10 eV in the sugar-phosphate volume produces such an effect. For the first decay the upper strand (which corresponds to the data in Figure 3) will break 12 bases to the right of the decay. In the second decay the break will occur 7 bases from the decay, and for the third no break will occur. It should be noted here that decays 2 and 3 have identical electron spectra and differ only in their initial directions and in the random numbers selecting the interactions. By choosing different values of the energy in the sugar-phosphate to break a strand, a distribution of lengths (12, 7, etc in the case above) corresponding to those in Figure 3 can be obtained. Values from 15 to 25 eV were examined and 17.5 eV was the best single value which fitted the experimental data for both sets of Auger electron spectra. This fit is also shown in Figure 3.

This technique of empirically fitting to the experimental data for strand breakage on a molecular level minimises possible problems which could be produced by using a water vapour code or the simple volume model.

STRAND BREAKAGE FROM INCORPORATED ^{125}I DECAYS

Method

Having established a criterion for single strand

breakage it is now possible to return to the data illustrated in Table 1 and examine the energy deposition patterns for the production of single and double strand breaks. Here it will be assumed that single strand breaks on opposite strands will produce double strand breaks. Chatterjee and Magee used a maximum separation of 10-11 nucleotides while Ito used approximately 15. Van Touw et al., (1985) have measured the maximum separation to be about 30 at normal temperatures in the phage ΦX174. This separation may depend upon the form of the DNA as discussed earlier. In this case, however, almost all of the single strand breaks which can combine are well within these distances and the calculation is not sensitive to the choice of maximum separation.

From Table 1 and for the first decay, the upper and lower strands will break in several places producing a double strand break with a short segment of DNA deleted. Decay number 2 will also produce a double strand break without perhaps the loss of nucleotide pairs. The third decay will give rise to a single strand break.

The analysis described above was carried out for 10,000 decays (each decay cycled 10 times) for both electron spectra. For each decay a decision was made on the type of break produced and in addition the total energy deposited in the molecule was recorded. Two types of results were produced; the average effect of the decay (the number of double strand breaks per decay for example), and a differential output relating total energy and type of break.

Average effects of the decay

As expected, the two electron spectra gave different numbers of double strand breaks per decay. The original source (with the greater number of electrons) gave 0.90 dsb/decay while the later one gave 0.65 dsb/decay. These data are for the number of double strand breaks produced near the site of the decay and do not include the possible effects of the longer range radiation. This contribution can be estimated from experimental data as follows; Blocher (1982) and Radford and Hodgson, (1985) have measured the number of double strand breaks/Gy/Dalton for mammalian cells exposed to x-rays to be about 8×10^{-12}. For ^{125}I the electron flux in the nucleus is also expected to be of low LET so that this figure can be used here. For a mammalian cell nucleus of 8 μm diameter containing 6 pg of DNA the dose per decay is 0.006 Gy (Schneeweiss et al., 1985). The nucleus will contain 3.6×10^{12} Daltons of DNA which then gives $8 \times 10^{-12} \times 3.6 \times 10^{12} \times 0.006 = 0.17$ dsb/decay from this component of the radiation. The total number of double strand breaks/decay now becomes 0.90 + 0.17 = 1.07 for one of the electron spectra and 0.65 + 0.17 = 0.82 for the other. Both of these values are in good agreement with the usually cited value of one.

It is worth noting here that the small OER exhibited by the iodine decay may be a consequence of this low LET contribution to the radiation effect. Using an OER of 3 for the production of dsbs for low LET radiation and 1 for large energy depositions gives (0.17 x 3 + 0.9)/1.07 = 1.3 as the OER for one electron spectrum and 1.4 for the other. These values, albeit for the initial yield of double strand breaks, agree with measured values for cell death (Koch and Burki, 1975; Schneeweiss et al., 1985).

An RBE for the production of double strand breaks can also be calculated from the data above. Here the total number of strand breaks per decay can be compared with the number produced by the same dose of low LET radiation. These figures are 1.07/0.17 and 0.82/0.17 or an RBE between 6.3 and 4.8.

Differential effects of the decay

In compiling the data, the type of break and the total energy deposited in the DNA were correlated. From these data the probability of producing an effect (no break, ssb, dsb) for a particular energy deposited in the DNA can be established. The raw data for 10,000 decays for the original spectrum is shown in Table 2. The table is organised as follows; the second column gives the number of decays depositing energy in the DNA in the interval shown. Thus in the first row 25/10,000 decays deposited between 0

Table 2. Correlation between total energy deposited in the DNA and the effect.

Energy interval (eV)	Frequency	No break	ssb	dsb
0 - 50	25	20	5	0
50 - 100	72	32	36	4
100 - 150	118	22	63	33
150 - 200	230	16	116	98
200 - 250	374	6	144	224
250 - 300	538	9	141	388
300 - 350	730	2	123	605
350 - 400	826	1	90	735
400 - 450	986	1	86	899
> 450	6101	0	107	5994

and 50 eV in the DNA, and of these, 20 produced no break and 5 produced a single strand break. Later, of the 538 decays depositing between 250 to 300 eV in the DNA, 9 decays produced no break, 141 produced ssbs, and 388 produced dsbs. From these latter data then 9/538 (1.7%) of energy depositions between 250 and 300 eV produce no break. For the other decays 141/538 (26%) gave ssbs and 388/538 (72%) lead to dsbs. More details on this calculation are given in Charlton and Humm, (1987).

The two sources of electron spectra gave essentially the same results which are shown in Figure 4.

The graph shows, for example, that if between 0 and 50 eV is deposited in the DNA, 80% of the depositions will not produce a break and the remaining 20% of the decays will produce a single strand break. If a decay deposits between 150 and 200 eV in the DNA roughly half of the decays will produce a single strand break and half a double strand break. As the energy deposited in the DNA increases so the fraction of decays producing dsbs increases. The figure supports the idea of a threshold energy deposition for radiobiological action as suggested by Goodhead et al., (1980) in that energy depositions below 100 eV produce few

Figure 4. The production of breaks of various types as a function of the total energy deposited in the DNA.

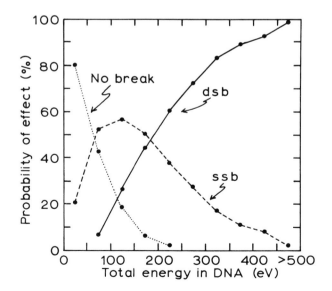

double strand breaks thought to be important initiating events in cell death. The threshold proposed by Goodhead et al. was for low LET radiation quite unlike that here.

CONCLUSIONS

A method of estimating the initial yield of single and double strand breaks near the decay of ^{125}I has been described. The results for average effects appear to be in good agreement with average values in the literature. This may be expected to produce about one dsb/decay. A more sensitive test of this method of calculation would be the initial yield of double strand breaks produced from a smaller electron cascade.

The work also demonstrates that the production of breaks is not a simple function of the energy deposited in the DNA but rather that for a particular deposited energy several different outcomes are possible.

REFERENCES

Blocher, D., 1982, International Journal of Radiation Biology, 42, 317.
Carlson, T.A. and White, R.M., 1963, Journal of Chemical Physics, 38, 2930.
Charlton, D.E. and Booz, J., 1981, Radiation Research, 87, 10.
Charlton, D.E. and Humm, J.L., 1987, (submitted for publication).
Chatterjee, A. and Magee, J.L., 1985, Radiation Protection Dosimetry, 13, 137.
Coghlin, W.A. and Clausing, R.E., 1973, Atomic Data, 5, 317.
Feinendegen, L.E., 1975, Radiation Environmental Biophysics, 12, 85.
Goodhead, D.T., Munson, R.T., Thacker, J. and Cox, R., 1980, International Journal of Radiation Biology, 37, 135.
Hofer, K.G., 1980, Bulletin of Cancer (Paris), 67, 343.
Humm, J.L., 1984, Ph.D. Thesis, KFA Report JUL-1932.
Ito, A., 1987, Advisory Group Meeting on "Nuclear and Atomic Data for Radiation Therapy and Related Radiobiology" (in press).
Kassis, A.I., Sastry, K.S.R. and Adelstein, S.J., 1987, Radiation Research, 109, 78.
Koch, C.A. and Burki, H.J., 1975, International Journal of Radiation Biology, 28, 417.
Kornberg, A., 1974, DNA Synthesis (W.H. Freeman)
Linz, U. and Stöcklin, A., 1985, Radiation Research, 101, 262.

Martin, R.F. and Haseltine, W.A., 1981, Science, 213, 896.
Paretzke, H.G., 1987, Radiation Track Structure Theory. In Kinetics of Inhomogeneous Processes, edited by G.R. Freeman (J. Wiley) p. 89.
Pomplun, E., 1987, Ph.D. Thesis, University of Dusseldorf.
Pomplun, E., Booz, J. and Charlton, D.E., 1987, (to be published).
Schneeweiss, F.H.A., Myers, D.K., Tisljar-Lentulis, G. and Feinendegen, L.E., 1985, Radiation Protection Dosimetry, 13, 237.
Radford, I.R. and Hodgson, G.S., 1985, International Journal of Radiation Biology, 48, 555.
Van Touw, J.H., Verberne, J.B., Retel, J. and Loman, H., 1985, International Journal of Radiation Biology, 48, 567.

THE POSSIBLE ROLE OF SOLITONS IN ENERGY TRANSFER IN DNA: THE RELEVANCE OF STUDIES WITH AUGER EMITTERS

K.F. Baverstock and R.B. Cundall

MRC Radiobiology Unit, Chilton, Didcot,
Oxon. OX11 ORD, U.K.

ABSTRACT

The interpretation of some experiments in which ionising energy is directly absorbed by DNA involve postulates that large scale energy migration takes place in DNA over long distances (kilobase pairs). A possible mechanism for such processes is provided by solitary vibrational waves called solitons. Solitons arise when a pseudo-one dimensional system with non-linear characteristics suffers a large local transitory displacement from equilibrium. Various synthetic polymers, such as polyacetylene are known, to sustain solitons and various physical properties of biopolymers such as DNA can be described in terms of 'open states' associated with inplane rotation of a group of the hydrogen bonded bases which has solitonic properties. The absorption of ionising energy by DNA systems can provide the transient displacement from equilibrium necessary to set-up wave conditions appropriate for soliton production.

Auger emitters would be particularly well suited for inducing solitons and offer the possibility for causing ionising energy to be 'injected' into the DNA molecule at a specific point in the molecular sequence. Experiments to test the hypothesis that this event causes long range energy transfer will be discussed.

INTRODUCTION

Despite extensive efforts over two or more decades a complete synthesis of the relationships between ionising energy deposition (i.e. track structure), molecular damage to DNA, and a biological endpoint has not yet been achieved even for the much studied effect of cell killing (see for example, Chadwick and Leenhouts, 1981). It may be argued that this measurable endpoint is not simple, in so far as it can involve damage to sensitive molecular structures other than DNA, eg. cell membranes. This is not the case for the mutation and cell transformation endpoints, or for that matter the cytogenetic endpoints which involve DNA alone. In none of these cases is there a complete mechanism known.

It may be that biological responses such as repair obscure the relationship between primary radiation processes and biological effects. In spite of the many suggestions as to how repair might intervene no underlying synthesis has been achieved.

The study of a part of the mechanistic synthesis, namely the interaction of radiation with DNA, has itself raised some interesting questions. An example is provided by the observations of Neary et al., (1972). They determined strand breakage in T7 bacteriophage DNA irradiated as a 'dry' film where radical diffusion would be reduced to a minimum, and the direct energy deposition mechanisms optimised. At high LET the number of single strand breaks per intersection of a track core with the DNA was as high as seven although it should be noted that this result depends on model for track structure employed. Also over a wide range of LET the ratio between double and single strand breaks was more-or-less constant. These results were regarded as suggesting a mediating role for the DNA between energy deposition and damage manifestation (Neary et al., 1972). This is an interesting suggestion which implies long-range (order of microns) energy excitation transfer. More conventionally the DNA is regarded as a passive entity in the radiobiological process and damage sites are spatially related to the pattern of energy depositions.

More recently one of us, Baverstock (1985), using a 'dry' film system similar to that used by Neary et al., (1972) has observed abnormalities in the distribution of DNA fragment lengths following irradiation. Utilising a technique for measuring the length distribution of the DNA fragments (Baverstock et al., 1983) it has been shown that for linear DNA (T7 bacteriophage) irradiation by low LET yields an excess of short fragments. Furthermore, when circular plasmid molecules are irradiated the excess of short fragments is seen to be matched by a deficit of linear molecules of length equal to the circle. The latter are fragments which have sustained just one double strand break.

We suggest that these results might be explained by the non-linear nature of the response to the process of deposition of ionising energy in the biopolymer, DNA, and that this results in the production of what are known as solitons. Solitons are solutions to non-linear differential equations which have clearly defined if, in terms of conventional linear physics, unusual properties. They are essentially one dimensional waves that maintain dynamic integrity by balancing the effects of non-linearity against those of dispersion (Scott et al., 1973). It has been postulated (Englander et al., 1981) that solitons can be propagated in DNA and are best visualised as mobile 'open states' in which a series of about 10 bases of the DNA have rotated to varying extents into an 'outward' position. Solitons must, by their nature, have velocities of less than that of sound in the same media. In ordered structures they travel with very little dissipation of their energy. They thus present a possible mechanism for the transfer of energy over substantial distances. Although solitons are vibrational in nature they can act as carriers of both excitons and electrons, see Davydov (1981).

In the context of excitation transfer in DNA we propose, that solitons provide mobile regions of the DNA which through 'an open state' support and transport electronically excited states (excitons) or charge (polarons). This would result in such excitons or electrons becoming mobile but restricted to motion of the distortion associated with the soliton. In this context we need to distinguish between highly 'ordered' structures, exemplified by crystals and 'soft' but 'organized' structures, typified by hydrogen bonded polymers such as nucleic acids and proteins. A high degree of 'order' is not necessary although coupling of vibrational modes of the DNA is essential. The DNA molecule can adopt a number of structural forms A, B, Z, etc. Structural discontinuities between these different structures could serve to bring about the dissipation of solitons giving rise to the 'dumping' of excitons or electrons and thus to strand breakage.

The conditions necessary for the formation of solitons in irradiated systems have been outlined by Bednár, (1985) and he points out that these conditions are fulfilled in particular by inner shell Auger transitions due to the energy released by Auger cascades. Since soliton formation is a non-linear effect the energization must be locally intense and removed from conditions of normal thermal equilibrium. Excitation by low intensity light cannot stimulate soliton formation because it involves violation by the Franck-Condon principle.

The purpose of this paper is to consider experiments using incorporated Auger emitters to assess the possible influence of DNA solitons and to review experiments that might be undertaken to examine the hypothesis further.

Experiments using positioned Auger emitters in DNA

Two sets of published experiments belong to this category, namely those of Martin and Haseltine, (1981) and also those of Linz and Stöcklin, (1985).

Martin and Haseltine, (1981) used linear DNA fragments of defined length (59 bp and 121 bases, only part double stranded) labelled with both ^{32}P and ^{125}I. A 59 bp fragment was labelled with ^{125}I at one end and ^{32}P at the other end of the same strand. Measurements of the length distribution of denatured fragments containing ^{32}P were used to assess how far from the ^{125}I label the strand was broken. The results indicated that most breaks occurred within a few bases of the ^{125}I residue. A longer fragment, partly double stranded, was labelled such that ^{125}I decay occurred at the end of the short strand. Breaks in the longer strand were measured.

Analysis of these experiments by Charlton and Humm, (1987) using theoretically generated electron tracks from ^{125}I Auger decays indicates that it is not necessary to invoke energy transfer in order to obtain the result observed by Martin and Haseltine, (1981).

In contrast Linz and Stöcklin, (1985) used pBr322 -a plasmid containing about 4000 bp, into which two ^{125}I residues were incorporated at a restriction enzyme cut site followed by, for some experiments, recircularisation (ligation) of the molecule. The treated plasmids were stored for three months in either sodium chloride solution (10 mM) or phosphate buffer (10 mM) both in the presence and absence of oxygen.

For a control the consequences of storing the plasmid unlabelled in the presence and absence of $Na^{125}I$ were compared. In the absence of the Auger isotope relatively little fragmentation occurred in the unligated (T^-) series, ie. linear plasmids. The presence of $Na^{125}I$, giving a dose of about 5 krad to the whole solution made little further effect. However, when the ^{125}I was incorporated into the unligated molecule extensive fragmentation occurred in both the sodium chloride solution and in phosphate buffer, there being more breakage in the buffer free solutions. The effect of the presence of oxygen was not marked in any of the experiments.

In the ligated DNA (the T^+ series in which 32% was present either as linear dimers or circular plasmids) the degree of fragmentation was reduced by a factor two. This reduction was mirrored in the yield of products associated with the destruction of the base containing the ^{125}I.

The authors attribute the differences in breakage between sodium chloride and phosphate buffer solutions to the radical scavenging properties of the phosphate buffer. This does not seem convincing in the light of their result obtained with Na^{125}I in solution. In this experiment most of the effect would be due to radicals generated from Auger electrons interacting with water remote in molecular terms from the plasmid molecules. Relatively few breaks were observed. In any case the primary effect of radical damage is to cause single strand breaks and these would only have been detected in the Linz and Stöcklin experiment when they were sufficiently numerous to lead to the formation of double strand breaks.

The authors discuss the possibility that fragmentation is the result of long range energy transfer. They propose that thymine residues might act as traps for excitational energy so precipitating breaks when the concentration of thymine residues is high. The problem with this interpretation is that on average 1 in 4 bases are thymine and so such traps are numerous in any DNA molecule of this type. It is true that 'thymine rich' zones exist and the authors examine the possibility that the fragmentation patterns they observe correspond to the distribution of such zones and they do adduce some evidence for this.

Comparison with other results

It is difficult to ignore the comparisons between this latter study using ^{125}I and the results from Neary et al., (1972) and from Baverstock, (1985).

The marked difference observed by Linz and Stöcklin, (1985) between the ligated and unligated samples both in terms of DNA fragment generation and in the generation of gaseous products reflects Neary's conclusions regarding the 'mediating' role of DNA between energy deposition and damage and the difference observed by Baverstock (1985) in the threshold energies required to cause a double strand break between T7 DNA and the plasmid. The suggestive evidence for more than one double strand break in the molecule per decay is precisely what is observed by Baverstock (1985).

As pioneering experiments both the Martin and Haseltine, and the Linz and Stöcklin studies represent major steps forward but each in its way has its limitations. Martin and Haseltine used a very short DNA fragment and as such were precluded from seeing long-range effects. The Linz and Stöcklin's experiment was confounded, to a degree, by the possibility of radical migration through water.

The Soliton Hypothesis

The conditions for the generation of a soliton in DNA, ie a sufficient displacement from equilibrium to involve passage through the non-linear boundary conditions are met both by high LET type radiation events and by Auger emission (Bednår, 1985). There is a proportion of high LET character even in so called low LET radiations, particularly at the ends of electron tracks. Thus we might assume that all ionising radiations are in principle capable of generating solitons although their effectiveness in doing so would be variable.

As is explained above, solitons are envisaged as carriers of excitation energy by virtue of the self propagating distortions they give rise to in the DNA helix. Energy localization is postulated to occur as the result of the collapse of the soliton or as the favourable coincidence of the excitation energy from two passing solitons. Trapping may occur as the result of a structural dislocation in the DNA. The sites of coincidence of excitation from colliding solitons may depend on the relationship between the site of deposition and points of reflection such as the ends of the molecular linear DNA. It should be emphasised that the energy of the soliton in linear DNA is not readily dissipated since an activation energy is required for this to occur (Davydov, 1981).

These are properties that give rise to the possibilities of testing the soliton hypothesis, in particular the capability using Auger emitters, to place the energy into the DNA molecule at a clearly defined site in relation to other features of the molecule, for example, the ends, a Z DNA inducing sequence, a palindromic sequence, a cross link, a bend around a histone like structure, etc.

Why invoke solitons?

There are two aspects to this question. Firstly, long-range energy transfer could offer an explanation of the results in some experiments (namely those of Neary et al., (1972) and Baverstock, (1985) in which ionising energy is directly deposited in DNA. Such a mechanism would imply either a very rapid transfer of excitation, such as might be envisaged if DNA was electronically delocalized or a slower transporting medium in which dissipation was minimal. We know from experimental measurements that native DNA in the ground state is not highly electrically conducting, (Whillans, 1975) and may not be sufficiently coupled electronically (Shapiro et al., 1975) for long distance excitation transfer to be likely. Evidence upon the latter remains a matter for debate however. The soliton satisfies the criterion of non-dissipation and so is an attractive possibility. Under linear conditions electronic excitation in DNA may have only a limited range (a few bases) of migration. However, under non-linear conditions, such as can be induced by ionising radiation this is not necessarily the case.

Secondly, it can be argued that solitons are precisely what would be expected to result from an ionising energy deposition into a DNA molecule. Biopolymers, such as DNA and proteins, have highly non-linear characteristics by virtue of their structure. The result of subjecting such structures to 'shocks' or other constraints which push them far from their equilibrium condition is to induce solitons or solitary vibrational waves.

A conceptual difficulty which arises is that given the context of classical and quantum physics as usually presented and the scientific framework which has grown up around them, solitons are what might be described as 'counter-intuitive'. The initial state generated by an ionising deposition would be typically described as 'chaotic'; rather in the sense that an explosion is seen as a 'chaotic' event. The detonation wave resulting from an explosion is however a highly ordered structure. In fact, such 'far from equilibrium' conditions can be the source of order (Prigogine and Stengers, 1984): a fact of considerable biological importance. A soliton is one such ordered structure. The properties, of coherent behaviour, and its non-dispersive nature, seem to 'buck' the laws of physics only when linearity is regarded as mandatory. We are now familiar with the generation of monochromatic and coherent light when a 'laser' cavity is 'super excited', another example of 'order out of chaos'.

The 'pebble in the lake' analogy of ionising energy deposition into DNA, where the waves decay rapidly as the distance from the event increases is not realistic. We suggest the better analogy would be the 'very large rock dropped into the very narrow canal'; similar circumstances to the first observation of a solitary wave or 'soliton' by John Scott Russell, in 1837 (Scott Russell, 1844).

Possible future experimental approaches using Auger emitters

The advantage that Auger emitters offer in this field is the ability to ensure an ionising event at a specified location in a DNA molecule although there is a large 'low LET' component of emission from, for example, ^{125}I. This was the unique pioneering step taken by Martin and Haseltine, (1981), and Linz and Stöcklin, (1985). However such experiments require considerable ingenuity in order to construct the DNA vehicles needed.

The first experiment to consider is a repeat of the Linz/Stöcklin experiment under conditions in which radical migration is eliminated and in which the recircularised plasmid is isolated from monomer and dimer linears. We plan to carry out this experiment with the 'dry film' target system (Baverstock, 1985). If, under these circumstances, ie., with the possibility of long range action by free radicals eliminated, we find a similar fragmentation to that observed by Linz and Stocklin (1985) and can eliminate the possibility that breaks might occur in the DNA remote from the ^{125}I simply due to the molecule 'crossing' at the incorporation point of the ^{125}I, then we will have made a very strong case for invoking some form of intramolecular action at a distance and thus long range energy transfer.

If a positive result is obtained an important variation will be to cut the recircularised plasmid at another point so that a linear molecule with the Auger emitters located away from the ends is generated. It is presumed that solitons are reflected at DNA ends and so this would be a way of investigating whether or not soliton/soliton interaction is important in energy trapping.

Further variations would be to use a gyrase to induce super coiling in the recircularised plasmid. This can induce specific structural features at palindromic sites.

There are also a variety of plasmids available with specific sequences incorporated, for example, pKH47 with a 50 bp run of A-T pairs and other plasmids with Z-DNA inducing sequences; the importance of these as trapping sites can therefore be investigated.

In the 'dry film' system, water content of the DNA can be controlled by controlling the relative humidity. Water content infuences structure and can be used to induce a transition from B-A form. The A-form of DNA is also induced if the DNA duplex is 'bent' sharply and such bending can be induced by precipitating the plasmid from low salt concentration with polyamines. Thus we have at our disposal a quite extensive range of techniques for manipulating DNA structure.

In the Martin and Haseltine experiments we consider that the fragments were not long enough to support a solitonic mechanism, if indeed, it is an acceptable hypothesis. Their elegant technique could reveal the processes that occur in the immediate vicinity of the Auger decay in a much longer molecule. The construction of a doubly labelled plasmid incorporating an Auger emitter at one end and the ^{32}P label some tens of base pairs down the molecule and just upstream of a suitable enzyme 'cut' site would be no easy task, yet the potential value of this experiment is such that it should be very carefully considered.

We have briefly reviewed above some of the many options open to experimental investigation and warranted if the soliton hypothesis is seriously considered.

IN CONCLUSION

While there is no absolutely compelling reason to invoke the soliton mechanism in the radiation chemistry of DNA the following facts have to be acknowledged:

1) Solitons are objects with a real existence and clearly defined properties that are propagated in polymers with non-linear characteristics.

2) DNA is one such non-linear polymer.

3) Ionising radiations, especially Auger electrons, have the properties thought likely to induce solitons (Bednár 1985).

4) There are anomalous results in DNA radiation chemistry briefly reviewed above which could be explained at least qualitatively in terms of a soliton mechanism (Neary et al., 1972; Baverstock, 1985).

REFERENCES

Baverstock, K.F. (1985). 'Abnormal distribution of double strand breaks in DNA after direct action of ionising energy.' International Journal of Radiation Biology, 47, 369-374.

Baverstock, K.F., Berriman, J., Parker, T. and Stephens, M.A. (1982). 'An improved technique of strand break analysis for isodisperse DNA.' Radiation and Environmental Biophysics, 21, 81-96.

Bednar, J. (1985). Electronic excitations in condensed biological matter. International Journal of Radiation Biology, 48, 147-166.

Chadwick, K.H. and Leenhants, H.P. (1981). The molecular theory of Radiation biology. Springer-Verlag Berlin.

Davydov, A.S. (1981). The role of solitons in the energy and electon transfer in one-dimensimal molecular systems. Physica 3D 1 and 2 1-22.

Linz, U. and Stöcklin, G. (1985). Chemical and biological consequences of the radioactive decay of Iodine-125 in plasmid DNA. Radiation Research, 101, 262-278.

Martin, R.F. and Haseltine, W.A. (1981). Range of radiation chemical damage to DNA with decay of ^{125}I.

Science, 213, 896-898.
Neary, G.J., Horgan, V.J., Bance, D.A. and Stretch, A. (1972). Further data on DNA strand breakage by various radiation qualities. International Journal of Radiation Biology, 22, 525-537.
Prigogine, I. and Stengers, I. (1984). 'Order out of chaos: man's new dialogue with nature', Heinemann London.
Scott, A.C., Chu, F.Y.F., McLaughlin, D.W. (1973). Proc IEEE 61 1443.
Scott-Russell. J. (1845). 'On Waves'. Report of the 14th Meeting of the British Association for the Advancement of Science, 13, 311-388.
Shapiro, S.L., Campillow, A.J., Kallman, V.H., and Good, W.B., (1975). Optics Comm., 15, 308-310
Whilans, D.W., (1975). Biochim Biophys. Acta, 414, 193-205 (1975).

DOUBLE STRAND BREAKAGE IN DNA PRODUCED BY THE PHOTOELECTRIC INTERACTION WITH INCORPORATED 'COLD' BROMINE

J.L. Humm[1] and D.E. Charlton[2]

[1] MRC Radiobiology Unit Chilton,
Didcot, Oxon. OX11 ORD

[2] Concordia University,
1455 de Maisonneuvue Blvd.,
Montreal, Quebec,
Canada H3G 1M8

ABSTRACT

Two papers in this book (Ohara et al. and Maezawa et al.) report on additional damage to irradiated cells containing the thymidine analogue bromodeoxyuridine (BrdU) incorporated into the DNA, when irradiated with monoenergetic x-rays just above the K-edge of bromine as against x-rays of energies just below the edge.

Using the track structure method described in the chapter by Charlton (ibid) the contribution to double strand break (dsb) production due to direct photoelectric interactions in bromine is calculated and compared with figures on initial dsb production from experiments with x- and γ-rays on different cell systems, unloaded with bromine, multiplied by experimental values on the enhanced dsb production caused by bromine sensitization. It is shown that the enhancement in initial dsb production in traversing the bromine K-edge is only 6%, even with a theoretical maximum loading of 100% replacement of thymidine with BrdU. Further, the enhanced effect in the comparison of bromine loading to non loading shows the dominant effect is due to the bromine sensitization of the DNA target. Finally, results are also given for the photoelectric effect in phosphorus and iodine incorporated into the DNA as iododeoxyuridine.

It is concluded that photoactivation of phosphorus, bromine or iodine is unlikely to be of potential value for therapy.

INTRODUCTION

Several groups (Halpern and Muetze 1978, Photon Factory Report 1987; Fairchild et al., 1982, for example) have reported the effects of the photoactivation of high atomic number atoms incorporated into biologically active molecules. The principle of this method is that by incorporating a target atom in the molecule, irradiation with photons of energy just above the K-shell binding edge of the atom will produce a selective absorption of the radiation and an effect elevated above that for photons of energy below the K-edge. This effect is predicated on the Auger cascade originating from the K-shell absorption being a major cause of the damage measured.

There are several other factors to take into account in the analysis of this problem however. The Auger cascades originating from K-shell vacancies are larger (on average) than L-shell vacancies only for atomic numbers less than 35 (Carlson et al., 1966). Above that, fluorescence radiation, that is the release of characteristic x-rays rather than Auger electrons, is a major pathway for the release of energy from the excited atom. Such a mechanism moves the K-shell vacancy to a higher orbital and considerably reduces the yield of Auger electrons. Secondly there will be an electron flux originating in the medium surrounding the molecule which will also contribute to its damage. In some cases the effect of this contribution is enhanced by the sensitization of the irradiated molecule by the incorporated target atom. This is particularly true in the case of bromine incorporated into DNA. It is likely that Auger cascade damage is not increased by this mechanism since the length of DNA effected by the cascade is small, and will not contain another bromine atom.

In this presentation the damage to DNA will be evaluated with and without incorporated bromine when irradiated by photons with energies at each side of the K-shell binding energy of the bromine (12.4 and 13.49 keV). The damage to the DNA will be calculated for the two sources discussed above, namely, from the induced Auger electron cascade and from the photoelectron flux generated outside the DNA. These are illustrated in the first figure.

Initial double strand breakage will be used as a measure of the damage. This is not necessarily directly proportional to cell death (Myers et al., 1977) a usual endpoint, but it provides a useful guide to the biological effect.

METHOD
Production of dsb from Auger cascades
The Auger cascade induced in bromine.

For the photon energies discussed here, just above and below the K-edges of bromine, the electromagnetic interactions are limited to the photoelectric effect.

Figure 1. The production of photoelectrons and Auger cascades near or in irradiated bromine loaded DNA

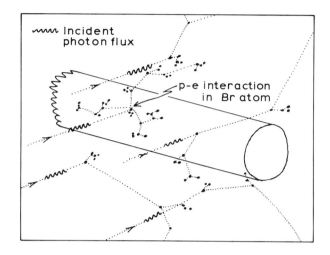

Table 1. Normalized initial inner shell vacancy distribution produced in bromine by photoelectric interaction (Scofield, 1973).

Shell/subshell	Below K-edge 12.4 keV	Above K-edge 13.49 keV
K	0.000	0.861
L_1	0.474	0.066
L_2	0.138	0.019
L_3	0.242	0.034
M_1	0.075	0.011
M_2	0.020	0.003
M_3	0.036	0.005
M_4	0.001	–
M_5	0.002	–
N_1	0.009	0.001
N_2	0.001	–
N_3	0.002	–

Table 1 shows the initial vacancies produced in the atom by these two radiations (Scofield 1973). Photons with energies above the K-edge predominantly remove one of the K-shell electrons. This does not necessarily lead to larger average cascades than those originating in the L-shell since the fluorescent yield for the K-shell is 0.618 (Krause, 1979) i.e. an electron falls from a higher sub-shell to fill the vacancy, radiating a characteristic x-ray rather than an Auger electron.

To generate the Auger electron spectrum the method described by Charlton and Booz, (1981) and Humm, (1984) was used. Here, using one of the photon energies, an initial vacancy was selected by a Monte Carlo method from the initial vacancy distribution, and the Auger cascade followed from the known transition probabilities (Chen et al., 1979; McGuire, 1972, 1975, 1977). One thousand individual spectra were generated for each photon energy and stored. Average values from these data are given in Table 2.

Table 2. Average quantities of the bromine Auger spectra per interaction and the effects in the DNA.

	Below K-edge 12.4 keV	Above K-edge 13.49 keV
No. of electrons	5.65	6.06
Total electron energy (keV)	12.30	6.96
Total energy in DNA (eV)	154	179
No. of dsb in DNA per cascade	0.34	0.41

Calculation of energy deposition in the DNA and the production of dsb per cascade.

The total energy deposited in the DNA for each cascade was calculated by modeling the DNA as a long cylinder of diameter 2.3nm with the bromine atom located at 0.15 nm from the central axis (its position when incorporated via the thymidine precursor BrdU). Each electron spectrum was taken in turn and using the electron track structure code MOCA7B (Paretzke, 1987) each electron was followed from its origin. The electron track code gives the energy deposited and the co-ordinates of each interaction, and it is straight forward to sum the energy deposited in the DNA volume for each cascade. The results of this calculation are given in Figure 2 which shows the energy deposition spectrum for the two photon energies for 10,000 cascades (1,000 photon interactions with each electron spectrum repeated 10 times by using new directions for the electrons each time) and in Table 2 in which the average energy per cascade is given.

In the figure the number of cascades depositing energy within various limits are given, for example, for photons above the K-edge, 2761 cascades deposited between 100 and 150 eV in the DNA. Note that cascades generated by photons with energy above the K-edge produce a greater number of energy depositions above 100 eV which, as will be seen later, are more effective at producing dsbs than energy depositions below 100 eV.

Figure 2. The energy deposition spectra in DNA from 10,000 cascades produced by photoelectric interactions in the bromine atom.

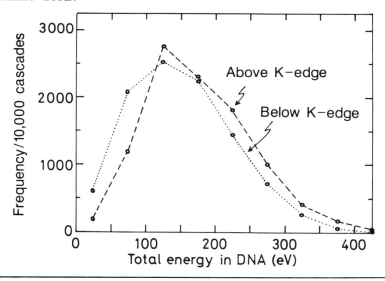

To calculate the production of double strand breaks from the energy depositions is more difficult. In the method used here a model of the DNA is required which allows the calculation of the energy deposited in volumes occupied by individual sugar-phosphate moieties along both chains of the duplex. Such a model is described in these proceedings (Charlton, 1987). For each cascade the energy deposited in these volumes was calculated. Previous work (Charlton and Humm, 1987) has shown that 17.5 eV in one of these volumes produces a single strand break. By examining the distribution of single strand breaks along each strand of the DNA a decision is made as to the consequence of each cascade. Three possibilities are evaluated - no break, a single strand break (ssb) or a double strand break (dsb). The technique is to assign a double strand break when single strand breaks on opposite strands are separated by less than 10 nucleotide pairs. Note that the effect of each cascade is evaluated which avoids the use of averages.

The results are shown in Figure 3 and Table 2. Figure 3 shows a general relationship between the energy deposited in the DNA and the effect produced. Cascades depositing

Figure 3. Probability of effect as a function of the total energy deposited in the DNA.

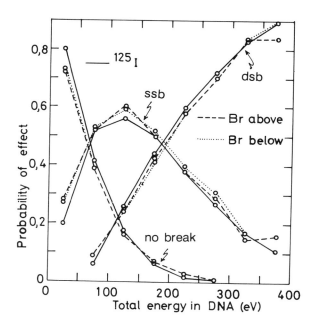

less than 100 eV produce few dsbs, the probability of which increases with increasing energy deposited. The data in Figures 2 and 3 may be combined to calculate the number of dsbs produced per cascade. As an example, 0.276 of the cascades initiated by photons above the K-edge deposited energy in the interval 100-150 eV (Figure 2), and the probability of this energy deposition producing a dsb is 0.24 (Figure 3). Thus 0.276x0.24=.066 dsb are produced for those cascades depositing between 100 to 150 eV in the DNA.

Summing over all of the cascades gives the figures shown in Table 2; 0.41 dsb/cascade for photons with energy above the K-edge and 0.34 dsb/cascade for those below.

In Figure 3 the data for previous calculations for incorporated ^{125}I (Charlton and Humm 1987) are included to demonstrate the apparent general relationship between the total energy deposited in the DNA by incorporated Auger emitters and strand breakage.

Table 3. Data used in calculating the number of incorporated bromine atoms hit per Gray.

	Below K-edge 12.4 keV	Above K-edge 13.49 keV
Mass energy absorption coefficient for water cm^2/g	2.63	1.92
Photons/cm^2-Gy	1.91×10^{11}	2.41×10^{11}
Photoelectric component of the attenuation coefficient for bromine cm^2/atom	3.33×10^{-21}	19.6×10^{-21}
No. of bromine atoms hit per Gray	$4.82 \times 10^{-13} \cdot N_m^1$	$3.58 \times 10^{-12} \cdot N_m$
No. of dsb/Gy from Auger cascades	$1.64 \times 10^{-13} \cdot N_m$	$1.47 \times 10^{-12} \cdot N_m$
No. of dsb/Gy from electron flux	$2.20 \times 10^{-11} \cdot N_m$	$2.20 \times 10^{-11} \cdot N_m$
Total No. of dsb/Gy	$2.22 \times 10^{-11} \cdot N_m$	$2.35 \times 10^{-11} \cdot N_m$

[1] N_m is the molecular weight of the genome in Daltons

Number of photoelectric interactions in bromine per Gray

The final step in the calculation of the number of dsbs produced by the Auger cascade is to evaluate the number of Auger cascades (bromine atoms hit) per Gray.

The number of photons/cm^2-Gy is given by $1/(h\upsilon \times \mu a/\varrho)$ where $h\upsilon$ is the energy of the photon and $\mu a/\varrho$ is the mass energy absorption coefficient for water (Hubbell,

1977). It is assumed that the volume surrounding the DNA (a sphere of radius in the order of the maximum range of the photoelectrons, about 3μm in this case) is sufficiently large that the bromine loading of the DNA does not affect the absorption coefficient. Values of μ_a/ρ and the photon fluence per Gy are given in Table 3 for the two photon energies used.

If N_m is the molecular weight of the DNA molecules in the target cell, then there are $N_m/660$ (660 is the molecular weight of an average nucleotide pair) pairs of bases in the cell. For 100% substitution of thymidine with bromodeoxyuridine approximately one quarter of all bases are labelled giving $N_m/2\times660$ bromine atoms in the genome. The number of bromine atoms in which the photoelectric effect occurs per genome per Gy is

$$\frac{N_m}{1320} \frac{\text{Br atoms}}{\text{genome}} \times \tau \frac{\text{cm}^2}{\text{Br atom}} \times \frac{1}{h\upsilon \times \mu_a/\rho} \frac{\text{photons}}{\text{cm}^2 \cdot \text{Gy}}$$

where τ_{Br} is the photoelectric component of the attenuation coefficient (Storm and Israel, 1970) and is given in Table 3. The expression above was evaluated and is also given in this table.

Each atom hit will produce an Auger electron cascade and earlier the number of dsb/cascade was derived. The product of these two quantities gives the number of dsb per Gy induced in the DNA genome by the Auger cascades. The results are given in Table 3.

Production of dsbs from the photoelectron flux

There have been several measurements of the production of dsbs by low LET radiation and these can be used to estimate the breakage of the DNA by the photoelectron flux. Frankenberg et al., (1986) for yeast report $0.8-0.5 \times 10^{-11}$ dsb/Gy/dalton, Bloecher, (1982) measured 1.2×10^{-11} dsb/Gy/dalton for Erhlich ascites tumour cells, and Radford and Hodgson, (1985) obtained a value of $0.6-0.9\times10^{-11}$ for V79 cells for doses greater than 6 Gy. It appears reasonable to use a figure of 1×10^{-11} dsb/Gy/dalton for the production of double strand breaks by this mechanism - independent of the photon energy.

The production of double strand breaks here will be increased due to sensitization of the DNA by the presence of bromine. Myers et al., (1977) have measured the increase in dsbs in T4 bacteriophage due to bromine loading to be 2.2. In this case 33% of the genome contained thymine moieties which had been replaced by 5-bromouracil so that this estimate may be high. It will be carried through the calculation and can be readily replaced when measured for the system being studied.

The number of double strand breaks/Gy from the photoelectrons will be $1 \times 10^{-11} \times N_m$ without sensitization and $2.2 \times 10^{-11} \times N_m$ with sensitization.

Bromine enhancement ratios

The contributions to double strand breakage from the two mechanisms of production discussed above are given in Table 3. The final results are that the enhancement produced in bromine loaded DNA by photons with energy above the K-edge compared to those of lower energy is 1.06. Comparing bromine loaded DNA with non-loaded DNA irradiated by photons with energy above the K-edge is $2.35 \times 10^{-11} / 1 \times 10^{-11}$ = 2.35, only slightly larger than the assumed sensitization produced by the bromine loading 2.2. Note that in these comparisons the enhancement is independant of the molecular weight of the genome.

It would appear from these calculations that damage produced by the induction of Auger cascades in bromine atoms incorporated into the DNA is unimportant compared to the effect of the sensitization of the DNA by the presence of the bromine. Clearly these results support the conclusions of Myers et al., (1977) and Majima et al., (1987) that selective K-shell ionisation of incorporated bromine atoms and the subsequent Auger electron cascade does not contribute significantly to the damage produced in cells loaded with bromine.

EFFECTS DUE TO AUGER CASCADES IN OTHER ATOMS

The calculation described above can be applied to other cold atoms in the DNA. Since there are experimental data available for incorporated iodine and for phosphorus, a brief examination of the effects of Auger cascades from these atoms will now be presented. All of the required data is available in the literature except for the new calculation of the number of double strand breaks per cascade which will be given for each irradiation condition.

Iodine as an incorporated atom

It has been suggested (Fairchild et al., 1982; Fairchild and Bond, 1984) that cold iodine when incorporated into DNA and irradiated with photons of energy above the K-edge (33.2 keV) will produce sufficient enhancement in addition to that of the sensitization to provide a useful mode of radiation therapy (Photon Activation Therapy, PAT).

To calculate the advantage gained from the induction of

Auger cascades in the iodine atom compared to irradiation without incorporated iodine requires the evaluation of the average number of dsbs produced by each Auger electron cascade. Using the methods described earlier and for photons above the K-edge gives 0.663 dsb/cascade.

Shinohara et al. (1987) have described an experiment of the irradiation of HeLa cells in which 20% of the thymidine was replaced by 5-iododeoxyuridine. The sensitization due to the incorporation was 1.8 as measured using ^{60}Co gamma-rays. For irradiation with sychrotron irradiation at 33.5 keV the dose enhancement ratio was 2.05.

For no incorporation, $1 \times 10^{-11} N_m$ dsb/Gy will be produced and at 33.5 keV irradiation with incorporated iodine $1.8 \times 10^{-11} N_m$ dsb/Gy are generated by the long range electron flux. At an energy of 33.5 keV there are 1.488×10^{12} photons/cm^2-Gy and the photoelectric attenuation coefficient is 7.41×10^{-21} cm^2/atom. For 20% replacement there will be $(0.2 \times N_m)/(4 \times 330)$ atoms of iodine in the genome. Combining these data with 0.663 dsb/cascade gives $1.11 \times 10^{-12} N_m$ dsb/Gy for the effect of the Auger cascades. The enhancement ratio for irradiation with and without the presence of iodine in the DNA is $(1.8+0.11)/1=1.91$, in good agreement with the measured value.

Since the value differs little from that for sensitization (1.8) it would appear that the enhancement predictions for the use of incorporated iodine for PAT are overly optimistic.

Auger cascades in phosphorus

Irradiation of DNA with photons at either side of the K absorption edge of phosphorus has been performed by Maezawa et al., (1987). Here the number of target atoms is equal to the number of bases and the complicating factor of sensitization is not present. Maezawa et al., 1987 obtained the result that there was no difference in effect for irradiation above and below the edge while at the edge there was a 1.5 enhancement of killing (at the 10% survival level) of D. radiodurans.

Calculation of the production of double strand breaks above and below the edge ($\tau = 1.37 \times 10^{-19}$ cm^2/atom above and 1.27×10^{-20} cm^2/atom below, 5.80×10^9 photons/cm^2-Gy) again requires the number of dsb/cascade. Placing the phosphorus atom near the edge of the DNA and using the technique described previously gave 0.228 dsb/cascade and 0.099 dsb/cascade for irradiation above and below the K-edge of phosphorus respectively. Using 1×10^{-11} dsb/Gy for the contribution of the electron flux to the damage gives an enhancement ratio of 1.05, in agreement with the experiment.

The effect at the edge may be due to resonance absorption in the oscillatory detail for the phosphorus atom which is not usually taken into account in published photoelectric cross-sections.

CONCLUSIONS

A method of calculating the production of double strand breaks has been described in detail and used to calculate the effect of simulating the Auger effect in atoms within the DNA. The ratios of effects calculated for several situations agree well with values measured from cell killing. It would appear that by far the largest effect produced by the incorporation of a foreign atom into the DNA is a sensitization of the DNA and the Auger cascades stimulated in this atom are relatively unimportant.

REFERENCES

Bloecher, D., 1982, International Journal of Radiation Biology, 42, 317.
Carlson, T.A., Hunt, W.E. & Krause, M.O., 1966, The Physical Review, 151, 41.
Charlton, D.E., 1987, (in this volume).
Charlton, D.E. & Booz, J., 1981, Radiation Research, 87, 10.
Charlton, D.E. & Humm, J.L., 1987, (submitted for publication).
Chen, M.H., Crasemann, B. & Mark, H., 1979, Atomic and Nuclear Data Tables, 24, 13.
Fairchild, R.G., Brill, A.B. & Ettinger, K.V., 1982, Investigative Radiology, 17, 407.
Fairchild, R.G. & Bond, V.P., 1984, Strahlentherapie, 160, 758.
Frankenberg, D., Goodhead, D.T., Frankenberg-Schwager, M., Harbich, R., Bance, D.A. & Wilkinson, R.E., 1986, International Journal of Radiation Biology, 50, 717.
Halpern, A. & Muetze, B., 1978, International Journal of Radiation Biology, 34, 67.
Hubbell, J.H., 1977, Radiation Research, 70, 58.
Humm, J.L., 1984, Ph.D. Thesis, KFA Report, JUL-1932.
Krause, M.O., 1979, Journal of Physical Chemical Reference Data, 8, 307.
Maezawa, H., Furusawa, Y., Hieda, K., Kobayashi, K., Suzuki, M., Mori, T. & Suzuki, K., 1987, Photon Factory Activity Report No.4, National Laboratory for High Energy Physics, KEK, Japan, p. 235.
Majima, H., Okada, S., Hieda, K., Kobayashi, K., Maezawa, H., Furusawa, Y., Yamada, T. & Ito, T., 1987, Photon

Factory Activity Report No.4, National Laboratory for High Energy Physics, KEK, Japan. p.136.
McGuire, E.I., 1972, Report, SC-RR-71-0835, Sandia Labs.
McGuire, E.I., 1975, Report, SC-RR-71-0443, Sandia Labs.
McGuire, E.I., 1977, Report, SC-RR-71-0075, Sandia Labs.
Myers, D.K., Childs, J.D. & Jones, A.R., 1977, Radiation Research, 69, 152.
Paretzke, H.G., 1987, Radiation Track Structure Theory. In Kinetics of Inhomogeneous Processes, edited by G.R. Freeman (J. Wiley) p.89.
Photon Factory Activity Report No.4, 1987, National Laboratory for High Energy Physics, KEK, Japan.
Radford, I.R. & Hodgson, G.S., 1985, International Journal of Radiation Biology, 48, 555.
Scofield, J.H., 1973, UCRL-51326.
Shinohara, K., Nakano, H., Hiraoka, T. & Ohara, H., 1987, Photon Factory Activity Report No.4, National Laboratory for High Energy Physics, KEK, Japan, p.278.
Storm, E. & Israel, H.I., 1970, Nuclear Data Tables A7, 565.

AN ADDITIONAL ENHANCEMENT IN BrdU-LABELLED CULTURED MAMMALIAN CELLS WITH MONOENERGETIC SYNCHROTRON RADIATION AT 0.09 nm : AUGER EFFECT IN MAMMALIAN CELLS

Hiroshi Ohara[1], Kunio Shinohara[2], Katsumi Kobayashi[3] and Takashi Ito[4]

1 Div. Physiol. & Pathol., Nat'l. Inst. Radiol. Sci.,
   4-9-1 Anagawa, Chiba 260, Japan.
2 Dept. Radiat. Res., Tokyo Metropol. Inst. Med. Sci.,
   3-18-22 Komagome, Bunkyo-ku, Tokyo 113, Japan.
3 Radiat. Biol., Nat'l. Lab. High Energy Phys. (KEK),
   1-1 Oho-machi, Tsukuba, Ibaraki 305, Japan.
4 Inst. Phys., Tokyo Univ., Meguroku, Tokyo 153, Japan.

ABSTRACT

Dose-dependent responses for cell killing (HeLa cells) as well as chromosome aberrations (CHO cells) were investigated in these two lines of cultured mammalian cells after unifilar labelling with 5-Bromodeoxyuridine BrdU (5×10^{-5}M) and following exposure to monoenergetic synchrotron radiations at wavelengths of 0.09 nm and 0.1 nm. The radiation effects were compared with those on unlabelled cells.

The radiation effects on unlabelled cells were not distinguishable, while the BrdU-labelled cells showed different responses against the two different radiations. The most pronounced sensitization for BrdU-labelled cells was induced by the radiation at 0.09 nm. The effects of 0.1 nm radiation were between those of unlabelled cells and labelled cells irradiated with 0.09 nm radiation. We consider that the additional enhancement in both cell killing and chromosome damage is possibly due to the induced Auger cascade in those BrdU-labelled mammalian cells by 0.09 nm monoenergetic radiation.

INTRODUCTION

Production of an electron vacancy in the K-shell of an atom results in the loss of several orbital electrons. Biologically, an inner-shell vacancy in an atom like iodine specifically incorporated into cellular DNA can be produced by photoelectric stimulation (Hofer & Hughes, 1971, Burki et al., 1973; Feinendegen, 1975; Bloomer & Adelstein, 1978; Commerford et al., 1980). The biological effects of the inner-shell vacancy have been mainly ascribed to local ionization density created by the emitted Auger electrons (Warters et al., 1977; Myers et al., 1977; Martin &

Haseltine, 1981; Kassis et al., 1982).

Our previous study (Shinohara et al., 1985) has shown that the radiation-induced cell killing effect in the BrdU-labelled HeLa cells in culture was greater when they were irradiated with monochromatic X-rays at 0.09 nm wavelength, slightly shorter than that for the K-absorption edge of bromine (0.092 nm). It was suggested that the observed increase in cell lethality was probably due to Auger cascades induced in the molecules of BrdU incorporated in cellular DNA. Observations of cell damage in structures such as chromosomes in nuclear DNA support this thesis. Here, the results of a chromosome study using a cultured mammalian cell line of Chinese hamster ovary (CHO) cells are reported. The experimental design is the same as that used in the cell killing study (Shinohara et al., 1985).

The results indicated that the production of chromosome aberrations by this monochromatic resonance radiation aimed at the substituted bromine atoms was enhanced more than that attributable to BrdU sensitization.

MATERIALS AND METHODS
Cells and Culture

A pseudo-diploid subline of Chinese hamster ovary (CHO) cells maintained in McCoy 5a medium supplemented with 15% fetal bovine serum (GIBCO, U.S.A.) and antibiotics, was selected for the present study. For the cellular irradiation, exponentially growing cells were grown on the surface of membrane filter discs (25 mm Fuji Photo Co. Ltd., Tokyo). Each cellular disc was placed in a culture dish (Falcon #3013) and incubated at 37^0C in 5% CO_2 and 95% air.

Substitution of BrdU

5-Bromodeoxyuridine (BrdU, Sigma, USA) and deoxycytidine (dC, Sigma Co. USA) were added to cells to obtain the final concentrations (5 X 10^{-5}M for BrdU and 1 X 10^{-5}M for dC), the cells incubated for 18 hrs, (corresponding to more than 1 generation time) and followed by 1 hr chase with fresh medium and finally rinsed twice to remove the excess amounts of BrdU and dC.

Irradiation with Monochromatic X-Rays

A cellular disc was placed in the centre of an irradiation plate and prevented from drying out by surrounding it with Dulbecco's PBS wetted blanket filter paper during the irradiation. Cellular discs were held vertically to the horizontal beam, the size of which was only 30 X 7 mm^2, without the lid on the plate. The beam was obtained, using the 111-channel cut Si crystal monochromator, from synchrotron radiation produced by electron storage ring at the Photon Factory, KEK, Tsukuba, Japan. The intensity of monochromatic radiation was monitored with a free air ionization chamber throughout the irradiation. The exposure dose was finally determined by the method of Hieda et al.,(1984).

Chromosome Analysis

Chromosomes were examined in mitotic cells arrested by colcemide treatment (10 ug/ml) during 8 hr of incubation after irradiation. The cells were collected by dispersion, washed once with PBS followed by hypotonic treatment with 0.75 M KCl solution for 20-30 min at 37^{0}C. The cells were then fixed in 2 changes of a mixture of ethanol and acetic acid (3 : 1), spread on clean slides in a humidified and temperature-controlled (35-37^{0}C) atmosphere, and air-dried. The slides were stained with 2% Giemsa solutions (Merck, F.R.G.). Metaphase figures to be scored were selected on these slides for quality and spread of constituent chromosomes. Isochromatid breaks and gaps, dicentrics, rings and dots were scored as the chromosome aberrations, while the chromatid type aberration were chromatid breaks, gaps and chromatid exchanges of various complexity. The total scores of each type of aberrations and their sum for each dose were averaged by the number of the cells observed, and the frequencies of aberrations were expressed on a per cell basis. In some quantitative analysis of cytogenetic data, aberrations except single and isochromatid gaps are expressed as chromosome breaks. Accordingly, single fragments and chromatid breaks were scored as one break, while those of dicentrics, rings, dots and isochromatid breaks were scored as two breaks. In chromatid exchanges, the number of breaks were determined by their configuration and the number of chromatids involved in the exchange figure.

Dose Response Analysis

A dose response function was fitted by the weighted least square method, allowing for non-constant variance, to the data on yield of chromosome aberrations. A linear quadratic function was selected for fitting the dose response data of chromosome aberrations and a linear function Y=aD+b for fitting chromosome breaks. The relative sensitizing ratio was estimated from a comparison of the dose response parameters of the constructed curves.

RESULTS

The CHO cell line showed a modal number of chromosomes of 2n=21 (range 20-22). The percentage of cells retaining this modal chromosome number is approximately 96%. There was, however, considerable variation in chromosome morphology from cell to cell, however, no cells showed acentric fragment-like chromosomes that were morphologically indistinguishable from radiation-induced acentric fragment chromosomes. The incidence of aberrations were 0.135 per cell (range 0.065-0.3) in a total of 1272 cells of unirradiated control cells.

Chromosome aberrations were scored in those mitotic cells that reached metaphase during 8 hrs post-irradiation incubation at 37^0C. The incidence of various types of aberrations were scored separately in the 4 different groups of irradiation experiments. The dose responses for the incidence of some specified types of aberrations were constructed. The comparison of dose response curves between the experiments are described in figure 1-4.

Figure 1 shows the dose response change in the frequency of total chromosome breaks/cell. The four curves constructed by using a linear function ($Y=aD+b$) vs. exposure dose showed pronounced difference in their sensitivity and slopes. The maximum effect was seen in brominated cells that were irradiated by X-rays at 0.09 nm, followed by brominated cells irradiated by X-rays at 0.1 nm. Unbrominated cells showed no significant difference in their response and were less sensitive than brominated cells. The two curves for unbrominated cells ran closer to each other. The difference in wavelength of the two monoenergetic X-rays appeared not to be meaningful.

Figures 2-4 showed the dose response changes in frequencies of aberrations that were identified as isochromatid gaps and breaks, rings and dicentrics, dots and a pair of acentric frangments. The data were fitted to a linear quadratic form of equation ($Y = aD + bD^2$). The results in Figure 2 indicate that the photons at 0.09 nm might have produced the maximum incidence and increase almost linearly with exposure dose, while the others showed a curve of upward concavity. The sensitization was found mostly the same as that in Figure 1. The unbrominated cells were again found to be less damaged.

The results in Figure 3 appear to be similar to the results in Figures 1-2. Isochromatid breaks showed quite similar tendency to those responses in Figure 1.

The result in Figure 4 shows that for the incidence of dicentrics and rings, frequently used as the cytogenetic index of lethal lesions, it was barely possible to

Figure 1   Dose response for the incidence of chromosome breaks, including single and multiple forms in 4 different combinations of treatments by BrdU pre-labelling and synchrotron radiations.   A:BrdU + 0.09 nm X-rays (Y=2.91(Gy)D+0.14.   B:BrdU + 0.1 nm X-rays (Y=2.79(Gy)D-0.97.   C and D:-BrdU + 0.9 (C) or 0.1 nm (D) X-rays (Y=1.95(Gy)D-1.2).

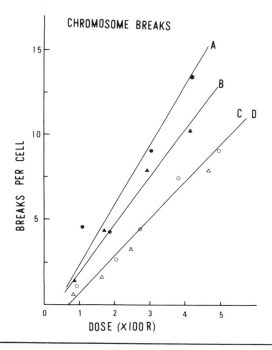

distinguish the difference between the brominated and unbrominated cells. However, no clear difference with wavelengths was seen between 0.09 nm and 0.1 nm in either the brominated or the unbrominated cells.

DISCUSSION

The aim of this study was to see if photoelectric stimulation, with appropriate energies of external radiation might cause inner-shell vacancies of electrons in incorporated bromine labelled DNA and thus increase cytogenetic damage in parallel with cell killing (Shinohara et al., 1985). If biologically active photoelectric resonance did occur the maximum effect would be expected for those BrdU-labelled cells that were irradiated by X-rays with wavelength just below K-absorption edge of bromine (0.092 nm).   Without BrdU-labelling, the cellular

Figure 2. Dose response for mult-hit types of aberration in 4 different combinations of treatment by BrdU pre-labelled and synchrotron radiations. Closed circles; BrdU+0.09 nm X-rays ($13.32 \times 10^{-3}D + 5.03 \times 10^{-6}D^2$): Closed triangles; BrdU + 0.1 nm X-rays ($Y=8.2 \times 10^{-3}D + 7.81 \times 10^{-6}D^2$): Open circles; 0.09 nm X-rays only ($Y=5.59 \times 10^{-3}D + 5.03 \times 10^{-6}D^2$): Open triangles; 0.1 nm X-rays only ($Y=3.52 \times 10^{-3}D + 1.67 \times 10^{-6}D^2$).

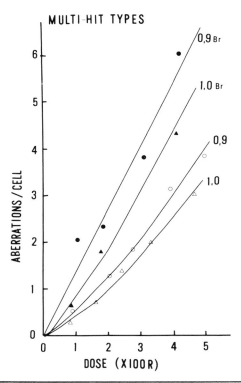

radiosensitivity may not be increased even if the energies of the two monoenergetic X-rays are different.

The biological effects of Auger electrons have been described in various kinds of biological systems when ^{125}I or ^{77}Br are incorporated into DNA (Hofer & Hughes, 1971; Burki et al., 1973; Koch et al., 1975; Kassis et al., 1982). From the microdosimetric point of view it is possible that such a photoelectric interaction may be enough to cause a double strand break in the DNA (Booz, 1984). The local energy depositions in DNA as well as in the cell nucleus as a result of the disintegration, or through critical photon interaction, are of interest in the

Figure 3 Dose response for incidence of isochromatid breaks in 4 different combinations of treatments by BrdU pre-labelling and synchrotron radiations. Closed circles; BrdU + 0.09 nm X-rays ($Y=8.57 \times 10^{-3}D + 2.17 \times 10^{-5}D^2$): Closed triangles; BrdU + 0.1 nm X-rays ($Y=5.54 \times 10^{-3}D + 3.85 \times 10^{-7}D^2$): Open circles; two kinds of radiations without BrdU ($Y= 3.52 \times 10^{-3}D + 1.67 \times 10^{-6}D^2$).

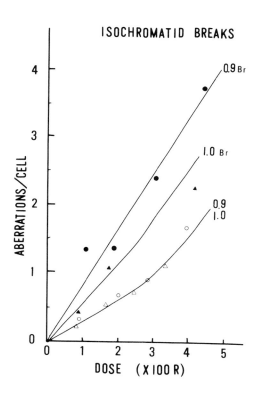

exploitation of Auger effect for radiation therapy (Fairchild et al., 1982; Bloomer et al., 1984; Bloomer et al., 1984; Fairchild and Bond, 1984).

The result in Figure 1 depicts the combined effects of the specific radiations and reactive chemicals on sensitization. The radiosensitivity of chromosome breaks/cell may be taken as the slope of the dose-response

Figure 4. Dose response for incidence of dicentrics and rings in 4 different combinations of treatments by BrdU pre-labelling and synchrotron radiations. Closed symbols are for the BrdU-labelled cells, and open symbols for unlabelled cells. Circles and triangles are for 0.09 nm and 0.1 nm X-rays, respectively. No fitting trials.

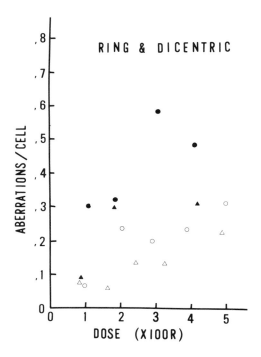

curves. The maximum radiation effect was found for the combined treatment with X-rays at 0.09 nm and BrdU-labelling, and there was some additional increase to the BrdU- sensitization levels with non-resonance X-rays at 0.1 nm. Without BrdU, there was no sensitivity difference. This can be interpreted as an indication of biological effect of Auger electrons emitted. In chromosome studies of radiation damage, it is said that there is some relation between the induction of reproductive death and multi-hit type of aberrations (Geard, 1983; Dewey et al., 1971; Davies and Evans, 1966; Schneider and Whitmore, 1963; Neary, 1965). The dose response of multi-hit type of aberrations resulting from low LET irradiation has been known to show an upward concavity, when the data are fitted either to a linear quadratic model or to the one track and two track model (Geard, 1983; Neary, 1965). The curves in Figures 3-5 were all of an upward concave type except for the one obtained with resonance radiation (0.09 nm) and BrdU-labelling; this was apparently linear even when fitted to linear quadratic model. Clearly, this linearity is mainly due to a large increment of aberrations for such low dose range below several Gray. Such a linear response has been observed for some of high LET radiation modalities (Chan et al., 1976; Sundell-Bergmann et al., 1985). Since high LET radiations can result in multiple induction of primary lesions due to a densely localized energy depostion, the formation of more complex and increased numbers of aberrations will give rise to the transformation of upward concavity into linearity. From microdosimetric considerations, it is likewise possible that Auger electrons may contribute to transformation of dose response curve by giving densely localized energy deposition in DNA, but their contribution at high doses may not be apparent. (Booz, 1984). In this context, the results in Figures 2-4 would suggest that Auger electrons may contribute to the formation of deletion type aberrations. Primarily, the formation of isochromatid breaks implies that two single lesions occur at an isolocus of two neighbouring daughter or unrelated chromatids. The Auger electrons emitted from the bromine molecules incorporated into DNA may happen to deposit a densely localized energy in very close proximity of other single lesions produced by BrdU and by the dose of X-irradiation, which may also come to produce a new pair of lesions in chromosomes. Then, the additional increment of isochromatid breaks with X-rays at 0.09 nm can be ascribed to the BrdU-labelling. This may be indicative of a possible biological effect of Auger electrons.

This interpretation may not apply to exchange types of aberration like rings and dicentrics. The results in Figure 4 in which a sensitizing effect with BrdU was clear but no differentiation of enhancement with wavelengths of radiation was observed and was not expected. Several possible explanations exist to account for the absence of the Auger effect. Firstly, the formation of rings and dicentrics specifically requires DNA synthesis after irradiation. Secondly, the distribution of cell cycle phases in each sample were not identical (there were variable fractions of S and G depending on irradiation dose). The samples consisted of those of cells which reached mitosis during 8 hrs of post- irradiation incubation. Thirdly, there are many reported results, in which the results are not deducible and often equivocal to look for a possible effect of Auger electrons (Myers et al., 1977; Maeda et al., 1986; Maezawa et al., 1985; Furusawa et al., 1986). The reasons for these contradictory results have not yet clarified.

Finally, on the present, as well as the previous results, we incline to the view that Auger electrons may induce some localized damage in chromatin structure, resulting in increments of cytogenetic lesions in DNA and chromosomes as well as cell death. But, more detailed consideration is required of the effect of radiation on the very initial phase of chemico-physical processes such as the possible contribution of either direct and indirect process to the formation of radiation induced biological damage.

Acknowledgement

We thank the staff members of Photon Factory at KEK, for their co-operation as well as for the understanding of radiobiological studies. This work was particularly supported by the grant aid of Japanese Ministry of Education for 1985.

References

Burki, H.J., Roots, R., Feinendegen, L.F. & Bond, V.P., 1973, International Journal of Radiation Biology, 24, 363.

Bloomer, W.D. & Adelstein, S.J., 1977, Current Topics in Radiation Research Quartery, 12, 513.

Bloomer, W.D., McLaughlin, W.H., Adelstein, S.L. & Wolf, A.P., 1984, Strahlentherapie, 160, 755.

Booz, J., 1984, Strahlentherapie, 160, 745.

Commerford, S.L., Bond, V.P., Cronkite, E.P. & Reincke, V., 1980, International Journal of Radiation Biology, 37, 547.

Chan, P.C., Lisco, E., Lisco, H., & Adelstein, S.J., 1976, Radiation Research, 67, 332.

Davies, D.R. & Evans, H.J., 1966, Advances in Radiation Biology, 2, 243.

Dewey, W.C., Stone, I.E., Miller, H.H. & Giblak, R.E., 1971, Radiation Research, 47, 672.

Diefallah, El-H. M., Stelter, L. & Diehn, B., 1970, Radiation Research, 44, 273.

Fairchild, R.G., Brill, A.B. & Ettinger, K.V., 1982, Investigative Radiology, 17, 407.

Fairchild, R.G. & Bond, V.P., 1984, Strahlentherapie, 160, 758.

Feinendegen, L.F., 1975, Radiation and Environmental Biophysics, 12, 85.
Furusawa, Y., Maezawa, H., Suzuki, K., Hieda, K., Kobayashi, K., Ito, K., 1986, Journal of Radiation Research, 27,17.
Geard, C. R., 1983, Effects of radiation on chromosomes. In Radiation Biology, edited by Pizzarello, D.J. and Colombetti, L.G., CRC press, 83.
Halpern, A. & Stocklin, G., 1974, Radiation Research, 58, 329.
Halpern, A. & Mütze, B., 1978, International, Journal of Radiation Biology, 34, 67.
Hieda, K., Kobayashi, K., Maezawa, H., Ito, T. & Ando, M., 1985, Photon Factory Activity Report, Nat'l. Lab. High Energy Phys. (KEK), 1983/84, p.VI-17.
Hofer, K.G., & Hughes, W.L., 1971, Radiation Research, 47, 94.
Hofer, K.G., Keough, G., & Smith, J.M., 1977, Current Topics in Radiation Research Quartery, 12, 335.
Kassis, A.I., Adelstein, S.J., Haydock, C., Sastry, K.S.R., McElvany, K.D. & Welch M.J., 1982, Radiation Research, 90, 362.
Koch, C.J. & Burki, H.J., 1975, International Journal of Radiation Biology, 28, 417.
Martin, R.M. & Haseltine, W.A., 1981, Science, 213, 896.
Maeda, I., Hieda, K., Maezawa, H., Kobayashi, K., Yamada, T., Ito, T., 1986, Journal of Radiation Research, 27, 17.
Maezawa, H., Furusawa, Y., Suzuki, K., Hieda, K., Kobayashi, K., Yamada, T. and Ito, T., 1985, Photon Factory Activity Report, 1984/85, 95 (VI-23).
Myers, D.K., Child, J.D. & Jones, A.R., 1977, Radiation Research, 69, 152.
Neary, G.J., 1965, International Journal Radiation Biology, 9, 477.
Schneider, O. & Whitmore, G.F., 1963, Radiation Research, 18, 286.
Shinohara, K., Ohara, H., Kobayashi, K., Maezawa, H., Hieda, K., Okada, S. & Ito, T., 1985, Journal of Radiation Research, 26, 334.
Sundell-Bergmann, S., Bergman, R. & Johanson, K.J., 1985, Mutation Research, 149, 257.
Warters, R.L., Hofer, K.G., Harris, C.R. & Smith, J.J., 1977, Current Topics in Radiation Research Quarterly, 12, 389.

EFFECTS OF AUGER CASCADES OF BROMINE INDUCED BY K-SHELL PHOTOIONIZATION ON PLASMID DNA, BACTERIOPHAGES, E.COLI AND YEAST CELLS

Hiroshi Maezawa[1], Kotaro Hieda[2], Katsumi Kobayashi[3] and Takashi Ito[4]

[1] Department of Radiation Oncology, Tokai University School of Medicine, Bohseidai, Isehara, Kanagawa 259-11, JAPAN
[2] Biophysics Laboratory, Department of Physics, Rikkyo University, Toshima-ku, Tokyo 171, JAPAN
[3] Photon Factory, National Laboratory for High Energy Physics, Oho, Ibaraki 305, JAPAN.
[4] Institute of Physics, College of Arts and Sciences, University of Tokyo, Meguro-ku, Tokyo 153, JAPAN

ABSTRACT

Several lines of study with DNA, phages, bacteria and mammalian cells have recently been performed to examine if an enhancing radiobiological effect of Auger cascade followed K-shell photoionization can be experimentally verified. Irradiation was carried out at Photon Factory with monoenergetic X-rays, just above and below the K-absorption edge (13.473 keV) of bromine. For monochromatization a channel-cut silicon crystal monochromator was used for the synchrotron radiation from the 2.5 GeV electron storage ring.

When bromouracil-labelled E.coli cells were irradiated with X-rays killing of cells was enhanced above the absorption edge, 13.49 keV, by 8% as compared with 12.40 keV (below the edge) only in the presence of 7.8% DMSO. In the case of dried BrdU-labelled T1 phage, a larger (about 26%) enhancing effect was observed. This would partly be due to the incomplete suppression of radical mediated process in E.coli cells. Various degrees of energy-dependent enhancement observed in the different biological systems are discussed both from the induced number of Auger events and from the increased energy absorption due to the presence of Br atoms in the system.

INTRODUCTION

It has long been suspected that innershell absorption of X-rays at a specified atom in a molecule could cause an effect to the molecule totally different from the ordinary type of absorption via ionization of outer orbit electrons (Platzman, 1952). A systematic attempt was made by Clark III in 1968 with Bacillus spores aiming at P K-shell ionization with fluorescence X-rays. As in earlier studies by Guild (1952), the attempt yielded no positive result that supports the occurrence of such a special effect. More recently, Halpern and Mütze, (1977) also attempted to show this with Micrococcus cells whose thymidines in DNA are partially substituted by BrdU, which gave a so-called Auger enhancement for inactivation due presumably to Auger cascades induced in the Br atoms.

X-ray sources used in these previous studies were of conventional solid target type. Therefore, numerous limitations associated with the source, such as low intensity and focusing problems, make it often technically not feasible to perform intended experiments precisely. Synchrotron radiation, with its unique properties overcomes these difficulties and, for the first time, it became possible to study the effects of photoinduced Auger cascade on the incorporated elements (artificially or naturally) in biological systems. It is distinctly different in its radiobiological purpose from that of the effects from internally administered radioactive isotopes undergoing K-capture decay (e.g., I-125). Our basic questions are concerned with what special molecular changes are induced compared to ordinary ionization and what would be their biological consequences (see review by Halpern, 1982).

This article summarizes our recent investigation on the biological effects of photoinduced K-shell ionization by Br atoms which are incorporated in DNA of plasmid, bacteriophage, E.coli, and yeast cells. With plasmid DNA a strand-break was measured, while inactivation was the observed effect for the other three materials. In all experiments, irradiation was carried out with monoenergetic X-rays precisely adjusted to below and above the K-absorption edge of Br atoms (13.473 keV) using synchrotron radiation combined with a crystal X-ray monochromator.

MATERIALS AND METHODS
pBR322 DNA

pBR322 plasmid DNA was isolated from E.coli HB101 and purified by CsCl density gradient (Maniatis et al. 1982). For preparation of 5-bromouracil (BrU)-incorporated pBR322 DNA, E.coli B/r (thy$^-$) cells carrying pBR322 plasmid were incubated first to OD 0.6 in Davis mineral medium supplemented with 20 mg BrU, 20 g casamino acids, 10 mg tryptophan, 10 mg methionine, 100 mg ampicillin and 10 mg tetracyclin per 1 litre and then for 16 hr with chloramphenicol (180 mg/l). BrU-incorporated DNA was purified in the same way as above. Substitution of thymine by BrU was determined as 97%, judged from the chromatogram of HPLC. One μl of DNA dissolved in buffer (3 mM Tris-HC1, 0.3 mM EDTA; pH 8.0) was dropped on a polypropylene sheet and dried in air. Irradiated DNA in solid was dissolved in distilled water and electrophoresed in 1% agarose (89 mM Tris-borate, 2 mM EDTA). Single-strand breaks were determined by a conversion of closed circular (form I) DNA to open circular (form II) DNA as described previously (Hieda et al., 1986).

Bacteriophage T1

Bacteriophage T1 was grown on E.coli strain B in a synthetic medium containing 200 mg/l 5-bromodeoxyuridine (BrdU) and 10 mg/l aminopterin. Phage particles were collected by centrifugation for 2 hr at 100,000 x g and resuspended in nutrient broth (8 g/l). Two μl of phage suspension was dropped on a small polypropylene sheet and dried in vacuo. After X-irradiation, phage particles were suspended in the 3 mM phosphate buffer containing 0.0025% gelatine. The surviving fraction was determined by the agar-layer method using E.coli strain B_{s-1} as an indicator.

E.coli cells

Method of incorporation of BrU into E.coli DNA was described elsewhere (Maezawa et al., 1987). In brief, E.coli cells (thymine requirement B/r strain) were precultured in Davis mineral medium supplemented with 2.5% casamino acids and 2 mg thymine. Exponentially growing cells were washed and incubated with 2 μg/ml BrU in Davis medium for 3 hr at $37^0 C$. Degree of BrU-incorporation was 87% as judged from the position of the peak in the CsCl density gradient. BrU-incorporated cells (1×10^8 cells) suspended in physiological saline with or without dimethylsulfoxide (DMSO) (7.8%) were placed on the sample area ($2-3 \times 5$ mm^2) of a Millipore membrane filter (pore size, 0.45 μm; diameter, 25 mm) for X-irradiation. The cell survival was determined from colony countings on nutrient broth agar plates.

Yeast cells

Diploid yeast (Saccharomyces cerevisiae) strain, D7T2, is a gift from Dr. B.J. Barclay. Cells were grown for 8 hr in YPD (yeast extract, peptone and dextrose) medium supplemented with aminopterin, sulfanilamide and BrdUMP to get BrdUMP incorporated. Harvested cells were suspended in 67 mM phosphate buffer and placed on a Millipore membrane filter for X-irradiation. The cell survival was determined by colony countings on synthetic complete agar plates.

X-ray irradiation

X-ray source

Synchrotron radiation (Photon Factory, National

Laboratory for High Energy Physics, Tsukuba) was used as a source of X-rays. Irradiation system of monoenergetic X-rays consists of a 111-channel cut silicon crystal monochromator and a sample chamber with a scanning stage (Kobayashi et al., 1987). The beam size of monoenergetic X-rays was 3 mm vertical and 33 mm horizontal. A specially designed free-air parallel plate ionization chamber was used for determining exposure. The exposure rate was about 3.5 kR/min at 12.4 keV.

Irradiation of plasmid DNA and bacteriophage T1

Plasmid DNA and bacteriophage T1 on a polypropylene sheet were irradiated with X-rays in a irradiation chamber through a mylar window (5 μm thickness). Nitrogen was continuously flushed during irradiation. The irradiation chamber was mounted on a scanning stage. Exposure rate was 1.8 kR/min at 12.4 keV in the scanning mode. Irradiation was performed at 12.40 keV and 13.55 keV for plasmid DNA and 12.40 keV and 13.51 keV for bacteriophage T1.

Irradiation of cells

Cells loaded on a membrane filter described above were mounted on a plastic dish with a moistened filter paper for irradiation. In order to get uniform irradiation, the cell-loaded plastic dish was moved up and down on a scanning stage according to a programmed schedule. In the scanning irradiation mode, exposure rate was 1.1 kR/min at 12.4 keV. Irradiation was performed at 12.40 keV and 13.49 keV for E. coli cells and 13.43 keV and 13.49 keV for yeast cells.

RESULTS AND DISCUSSION
Single-strand breaks of plasmid DNA

The remaining fraction of form I of pBR322 DNA decreased exponentially with increasing exposure (kR) at 12.40 keV and 13.55 keV energies with and without BrU (Figure 1). A small difference with 12.40 keV X-rays was observed regardless of the incorporation of Br. However, a significant sensitization was present in BrdU-DNA at 13.55 keV. The ratio of Do of normal DNA to that of BrU-incorporated DNA at 13.55 keV was 1.55 (Hieda et al., unpublished).

Inactivation of bacteriophage

Inactivation of bacteriophage T1 with and without BrdU-incorporated also decreased exponentially with increasing exposure (kR). BrdU-incorporated phages were generally sensitized to a greater extent than normal phages to both radiations. A larger sensitization was observed with 13.51 keV X-rays. The ratio of Do of normal phage to that of BrdU-incorporated phage at 13.51 keV was 2.07.

Figure 1. Single-strand breaks in irradiated pBR322 DNA by X-rays at 12.40 keV and 13.55 keV with (———) and without (- - -) BrdU.

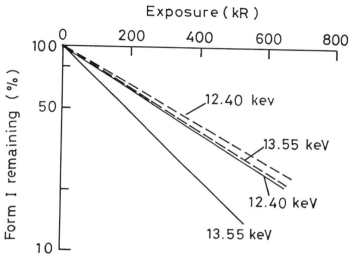

Figure 2. Inactivation of irradiated bacteriophage T1 by X-rays at 12.40 keV (circles) and 13.51 keV (triangles) with (———) and without (- - -) BrdU.

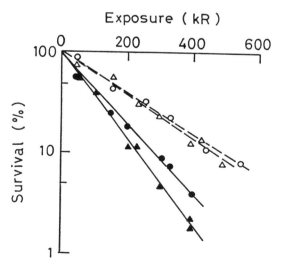

Inactivation of E.coli cells

In the case of E.coli cells irradiated with 12.40 keV and 13.49 keV radiations in saline as described above, Br-sensitization was significantly large, but independent of X-ray energy. When cells were irradiated in the presence of DMSO (radical scavenger), Br-sensitization was observed at both energies: 2.37 at 12.40 keV and 2.56 at 13.49 keV as calculated by the ratio of respective Do.

Inactivation of yeast cells

In yeast cells irradiated with 13.49 keV and 13.43 keV X-rays, a very large Br-sensitization was observed (Figure 3). Br-sensitization was 2.97 at 13.43 keV and 3.37 at 13.49 keV.

The above mentioned sensitization data are summarized in Table 1.

Auger enhancement in terms of the number of Auger events

Auger enhancement after K-shell photoionization of Br may be looked at in terms of the number of Auger events. We adopt a scheme depicted in Figure 4, in which non-specific sensitization by the presence of Br is taken on X axis (Br-sensitization) and apparent enhancement due to Auger event is taken on Y axis. Both are scaled (e.g., 10% increase of effect would be plotted as 1.1). The area on the graph may be regarded as the cross-section (reciprocal of the so-called Do or probability) of the action. At an energy above the K-absorption edge of Br atoms, energy-dependent enhancement (experimentally measurable) contains both apparent Auger enhancement and Br-sensitization occurring also at 12.40 keV or below the K-edge (see Figure 4c). The apparent Auger enhancement (E) is defined by Equation 1.

$$E = \frac{\text{sensitization at energy above K-edge}}{\text{sensitization at energy below K-edge}} \quad (1)$$

$$= \frac{[Do(-BrdU, 13.5 \text{ keV})]/[Do(+BrdU, 13.5 \text{ keV})]}{[Do(-BrdU, 12.4 \text{ keV})]/[Do(+BrdU, 12.4 \text{ keV})]}$$

It was 1.42 with pBR322 DNA, 1.26 with bacteriophage T1, 1.08 with E.coli cells and 1.14 with yeast cells (Table 1). The apparent Auger enhancement consists of true Auger enhancement (hatched area A in Figure 4) and Br-sensitization occurring at 13.5 keV (assumed to be equal to that for 12.40 keV). Under the assumption that non-specific sensitization by the presence of Br is equal for 12.4 keV and 13.5 keV X-rays, a virtual lethal dose (Doa) in terms of a true Auger event alone can be calculated from

Equation 2.

$$D_{oa} = \frac{E-1}{E} \times \frac{1}{\text{Br-sensitization at 12.4 keV}}^{-1} \times [D_o(+ \text{BrdU}, 13.5 \text{ keV})] \qquad (2)$$

Calculated values are listed in Table 1.

At the energy for the Br K-edge, the number (N) of Auger events in Br-incorporated DNA per unit exposure (kR) is calculated by Equation 3.

$$N = \Sigma n$$
$$= \Phi \sigma n \qquad (3)$$

where n is the number of Br atoms in the DNA, Σ the probability of photoionization in the K-shell per Br atom per unit exposure (kR), Φ the photon flux of X-ray energy above the K-absorption edge of Br and σ the photoionizing cross-section of the Br K-shell (Scofield, 1973). Calculated values for various materials are also listed in Table 1. Using values of Doa and N obtained above, the number of Auger events expected to occur per lethal exposure are calculated (last column in Table 1).

It can be seen from Table 1 that the number of Auger events per lethal exposure is grouped into two; one includes plasmid DNA and bacteriophages and the other cellular systems. It is to be noted that the latter group was irradiated under wet conditions, that is, DNA may have a large quantity of water among others in the close vicinity. This may reduce the probability that emitted Auger electrons actually hit the DNA. Also, part of the damage (e.g., single-strand break) could be repaired in cellular systems which may cause the decrease of efficiency for the observed effect. It is rather surprising that one Auger event produces 14 strand breaks in plasmid DNA, while corresponding efficiency for the inactivation with cellular systems seems not totally unreasonable.

Auger enhancement in terms of the quantity of energy aborbed

Effects of Auger cascade can also be discussed in terms of energy absorbed. That is, the enhancement of biological effects in Br-incorporated material after irradiation at X-ray energy above the Br K-edge could be a consequence of the increased dissipation of energy due to the absorption

Figure 3. Inactivation of BrdUMP incorporated (———) and normal (- - -) (+dTMP) yeast cells irradiated with 13.43 keV (closed symbols) and 13.49 keV (open symbols) X-rays.

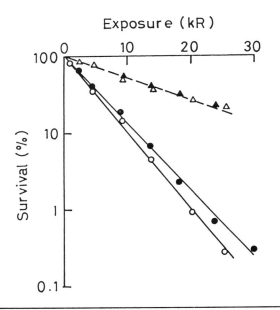

Figure 4. Diagrammatic representation for energy dependent enhancement of biological effects in the presence and absence of Br atoms. 'A' represents true Auger enhancement (see text for details).

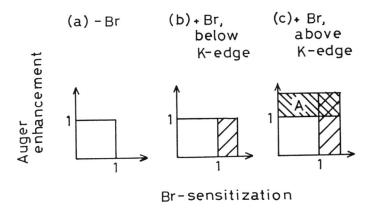

Table 1. Summary of data related to Auger enhancement of biological effects in Br-incorporated materials.

Materials	M.W. of DNA (dalton)	Degree (%) of substitution by Br	Auger event (N) in DNA per unit exposure (1/kR)	Br-sensitisation (at 12.4 keV)	Apparent Auger enhancement (E in Eq. 1)	Do (kR) for Br-incorporated (at 13.5 keV)	Lethal exposure[a] (Doa) (kR)	Auger event per lethal exposure[a]
pBR322 plasmid DNA	2.9×10^6	97	7.2×10^{-5}	1.09	1.42	271	999	0.072 (14)[b]
Bacteriophage T1	3.1×10^7	(50)[c]	4.3×10^{-4}	1.64	1.26	118	938	0.04 (2.5)[b]
E.coli cells[d]	5.3×10^9 (2 chromosomes)	87	0.13	2.37	1.08	5.57	178	23 (0.043)[b]
Yeast cells	1.8×10^{10} (diploid)	(50)[c]	0.25	2.97	1.14	5.86	142	36 (0.028)[b]
Mammalian cells[e] (HeLa)	1.2×10^{13}	(20)[c]	79	1.22	1.16	0.138	1.22	96 (0.010)[b]

a) Calculated exposure corresponding to hatched area A in Figure 4 b) Number of hits per Auger event (reciprocal of Auger event per lethal exposure). c) Assumed.
d) In the presence of DMSO. e) Shinohara, et al., (1985).

Table 2. Summary of data for absorbed energy and Br-sensitization.

Materials	Br	f-factor[a] (rad/R)		Do (Gy)		Br-sensitization in terms of absorbed energy	
		12.4 keV	13.5 keV	12.4 keV	13.5 keV	12.4 keV	13.5 keV
pBR322 plasmid DNA[b]	−	1.62	1.63	7440	6900		
	+	1.86	2.86	7840	7750	0.95	0.89
Bacteriophage T1[c]	−	0.949	0.966	2209	2376		
	+	0.949	0.966	1348	1140	1.64	2.08
E.coli cells[d]	−	1.17	1.17	167	167		
	+	1.18	1.19	71.9	66.3	2.32	2.52
Yeast cells[e]	−	0.987	0.988	194	195		
	+	0.990	0.996	65.7	58.4	2.95	3.34

a) Mass energy absorption coefficient of elements is taken from data of Hubbell (1982). f-factor = 0.870 $(\mu_{en}/\varrho)_m/(\mu_{en}/\varrho)_{air}$
b) Relative abundance of elements was calculated for DNA, Tris-HCl and EDTA system.
c) Relative abundance of elements was calculated for bacteriophage particles and nutrient broth system.
d) Relative abundance of elements was calculated using the chemical composition of cells (Watson, 1977) in the presence of DMSO.
e) Relative abundance of elements was assumed as the same as in E.coli cells.

by the incorporated Br atoms in accordance with its fractional amount.

From this point of view, Do in kR units in earlier sections was converted in Gy units using the R-rad conversion factor in the presence and absence of Br-incorporated in the systems for the two X-ray energies. The calculated Do in Gy are listed in Table 2 along with the conversion factors used. The sensitization based on the Do (in Gy) are shown in the last column in Table 2 for 12.4 keV and 13.5 keV, below and above the Br K-edge, respectively. From the difference between Br-sensitizations at the two X-ray energies it can be seen that, with the exception of plasmid DNA, where energy dependent enhancement (Figure 1) is explicable simply in terms of the increase of absorbed energy in the system, a significant degree of enhancement (at 13.5 keV) in other systems cannot be explained by the increase of energy absorbed in the system. This may mean that there may be a special effect associated with the mode of energy deposition by Auger electrons. In other words, for bacteriophage T1, E.coli cells and yeast cells the damage seems not to occur in proportion to the energy absorbed when Br is present in the system; one might envisage a certain lesion with a reduced repairability as compared with the lesion induced by the energy deposition of ordinary structure.

Although slightly different energies have been used in the experiments as the energy for below and above K-absorption edge of Br atoms they are represented by 12.4 and 13.5 keV, respectively, in the discussion of results and tables for the simplicity.

REFERENCES

Clark III, B.C., 1968, Ph.D. Thesis, (Columbia University).
Guild, W.R., 1952, Archives Biochemistry and Biophysics, 40, 402.
Halpern, A., 1982, In Uses of Synchrotron Radiation in Biology, edited by H.B. Stuhrmann (Academic Press, London), p. 255.
Halpern, A. & Mütze, B., 1977, International Journal of Radiation Biology, 34, 67.
Hieda, K., Hayakawa, Y., Ito, A., Kobayashi, K. & Ito, T., 1986, Photochemistry and Photobiology, 44, 379.
Hubbell, J.H., 1982, International Journal of Applied Radiation and Isotopes, 33, 1269.
Kobayashi, K., Hieda, K., Maezawa, H., Ando, M. & Ito, T., 1987, Journal of Radiation Research, (submitted).
Maezawa, H., Hieda, K., Kobayashi, K., Furusawa, Y., Mori,

T., Suzuki, K. & Ito, T., 1987, *International Journal of Radiation Biology*, (submitted).

Maniatis, T., Fritsch, E.F. & Sambrook, J., 1982, *Molecular Cloning*, (Cold Spring Harbor Laboratory, New York).

Platzman, R.L., 1952, In *Symposium on Radiobiology*, edited by J.J. Nickson (Wiley, New York), p. 97.

Scofield, J.H., 1973, *Theoretical photoionization cross sections from 1 to 1500 keV*, (UCRL-51326, Lawrence Livermore Laboratory, Livermore California).

Shinohara, K., Ohara, H., Kobayashi, K., Maezawa, H., Hieda, K., Okada, S. & Ito, T., 1985, *Journal of Radiation Research*, 26, 334.

Watson, J.D., 1977, *Molecular Biology of the Gene*, 4th edn. (Benjamin, California), p. 69.

RADIOTOXIC EFFECTS OF ^{125}I-LABELLED THYROID HORMONES WITH AFFINITY TO CELLULAR CHROMATIN

S. Sundell-Bergman, K.J. Johanson & G. Ludwikow

Department of Radioecology
Swedish University of Agricultural Sciences
S-750 07 Uppsala, Sweden

ABSTRACT

The thyroid hormones - thyroxine (T4) and triiodothyronine (T3) - have a profound effect on growth, development and metabolism of essentially all tissues in higher organisms. T4 and T3 are distributed to various target organs within the body where they are bound to the cellular chromatin via a specific nuclear receptor, a nonhistone protein. Abundant evidence supports the concept that most of the significant cellular responses to thyroid hormones in mammalian cells are mediated by a receptor localized to the chromatin of the target cells. T4 derives most of its biological activity through transformation to T3 which is the most active hormone and which is more readily bound to the receptor in the nucleus. The number of receptors per cell varies between different kinds of cells. The anterior pituitary, where T3 regulates the growth hormone gene transcription, show the highest frequency of specific T3-receptors.

Radiotoxic studies have been performed on cells in vitro after labelling with ^{125}I-T3. It is well established that ^{125}I displays a remarkable radiotoxicity when incorporated into DNA as iododeoxyuridine (IdU). We have investigated the induction and rejoining of DNA strand breaks in mammalian cells after uptake of ^{125}I-T3. Our results reveal a dose dependent increase in the number of breaks induced by ^{125}I-T3. In GH1 cells we found that between 1 and 3 unrepaired strand breaks were induced per ^{125}I decay. This finding is in close agreement with earlier results obtained with ^{125}IdU. Studies on micronuclei formation in GC cells after ^{125}I-T3 labelling show an accumulation of micronuclei with dose.

INTRODUCTION

The thyroid hormones - thyroxine (T4) and triiodothyronine (T3) - exert profound effects on growth, development and metabolism of essentially all tissues in higher organisms (Samuels, 1983). The hormones bind to high-affinity, chromatin-linked receptors in various target cells (Oppenheimer, 1979). T4 derives most of its biological activity through transformation to T3 which is the most active hormone and more readily bound to the receptor in the nucleus (Crantz et al., 1982). Several lines of evidence indicate that this binding plays a significant role in eliciting thyroid hormone effects (Oppenheimer et al., 1973; Yaffe & Samuels, 1984). Tissues known to be responsive to thyroid hormone are e.g. the anterior pituitary and the liver, which contain high concentrations of nuclear T3 receptors (Oppenheimer et al., 1974). In non-responsive tissues such as the spleen and the testis, the concentration of nuclear T3 receptors is low. Obviously there is a high degree of correlation between the number of nuclear receptors and tissue response (Oppenheimer et al., 1973).

The hormone receptors have been characterized in various tissues of rat (Oppenheimer et al., 1974) and human (Bernal et al., 1978). Cumulative evidence, such as binding of ^{125}I-T3, suggests that the protein encoded by the gene C-erb-A is identical with the thyroid hormone receptor (Sap et al., 1986; Weinberger et al., 1986). The receptor-chromatin interaction is poorly understood. It is not known whether the T3 receptors are homogeneously organized in chromatin or localized to restricted chromatin domains.

Evolutionarily the thyroid hormones have assumed a dual role - primarily in the development of the vertebrates and secondarily in the regulation of energy conversion (Weirich et al., 1987). T3 plays key roles in the regulation of the differentiation processes of the vertebrates while its regulation of temperature adaptation and calorigenesis has evolutionarily been a later phenomena. Most of the actions by thyroid hormones are initiated by binding to specific chromatin-associated receptors in nuclei of target cells.

The high radiotoxicity of ^{125}I incorporated into DNA as iododeoxyuridine (IdU) is well established and is mainly due to highly localized energy deposition by numerous electrons close to the decaying atom in DNA (Charlton, 1986). This extreme radiotoxicity has been manifested by measurements of DNA strand breaks (Sundell-Bergman & Johanson, 1980), chromosomal aberrations (Chan et al., 1976; Sundell-Bergman et al., 1985), cell survival (Bradley et al., 1975; Burki et al., 1973) and mutation induction

(Gibbs et al., 1987; Liber et al., 1983).

^{125}I-labelled T3 is not distributed uniformly in the nucleus but bound to a limited number of specific receptors followed by association with DNA. Evidently this process, in contrast to ^{125}IdU, limits the nuclear uptake of ^{125}I by the saturation of these nuclear receptors. The T3 receptor has been estimated to have a molecular weight of approximately 50 kilodaltons and a Stokes radius of 3.5 nm (Aprilletti et al., 1983). The exact localization of the hormone in relation to DNA is not known but the close kinship to the steroid receptor family would suggest that the hormone binding is not in the DNA binding region of the receptor (Weinberger et al., 1986). The structure of the receptor with bound T3 is not known but it would not be too far from reality to suggest that DNA may be in the high-LET sphere of the ^{125}I-decay (Figure 1). Our studies on the accumulation of DNA strand breaks indicate that T3 is located very close to DNA since the effects of ^{125}I-T3 and ^{125}IdU are similar.

We have studied the accumulation of DNA strand breaks in Chinese hamster cells (CHO, Cl.1, V79) and rat pituitary tumour cells (GH1, GC) after administration of ^{125}I-labelled T3 (Sundell-Bergman & Johanson, 1982, 1983). Pituitary cells are advantageous because of the high number of T3 receptors. Chromosomal damage has been measured by monitoring the frequency of micronuclei in GC cells. Micronuclei are supposed to be formed mainly from ascentric chromosomal fragments and show a close relation to chromosomal aberrations (Schmid, 1975).

MATERIAL AND METHODS
Cell culture

Growth of CHO and GH1 cells followed the methods described earlier (Sundell-Bergman & Johanson, 1982, 1983). GC cells (obtained from Dr. H. Samuels, New York) were grown at 37°C (5% CO_2 - 95% air) in Dulbecco's medium supplemented with 10% fetal calf serum (Flow. Labs.) and antibiotics. The cells were routinely checked for mycoplasma infection.

Radioactive labelling

Labelling of cells for DNA strand break analysis has previously been described (Sundell-Bergman & Johanson, 1982).

GC cells, used in micronuclei experiments, were labelled with L-3,5,3'-^{125}I-triiodothyronine (spec. activity approx. 45 MBq/ug, NEN) at various radioactive concentrations (3.7 to 37 kBq/ml). After a labelling period of 72, 96 or 144 hr at 37°C the cells were washed, nonradioactive medium was added and the incubation continued for a further 0, 18 or 27 hr.

^{125}I-T3 uptake

To study cellular versus nuclear uptake, exponentially growing CHO and GC cells were incubated in serum depleted medium at various radioactive concentrations of ^{125}I-T3 at 30°C. After 24 hr the cells were washed twice in medium and once in phosphate buffered saline (PBS), then scraped and suspended in PBS. Before centrifugation, samples were taken for determination of cell density (Coulter counter) and ^{125}I-activity (NaI-detector). The cell pellet was suspended in buffer A (10 mM Tris, 3 mM $CaCl_2$, 2 mM Mg-acetate, 0.5 mM dithiothreitol, pH 8.0) and after 5 min on ice, 0.25 ml of 10% Triton X-100 was added per millilitre of cell suspension. The cells were homogenized for 10 strokes in a glass homogenizer with a Teflon pestle. The nuclei were pelleted by centrifugation at 1100 g for 10 min, washed and suspended in buffer B (10 mM Tris, 0.25 M sucrose, 3 mM $MgCl_2$, pH 7.4). Finally the nuclei concentration and the ^{125}I-activity were estimated.

DNA strand breaks analysis

The number of DNA strand breaks was estimated by the DNA unwinding technique in combination with hydroxylapatite chromatography. The method is described in detail elsewhere (Rydberg, 1975).

Micronuclei assay

Cells were centrifuged at 3000 rpm for 40 sec. The pellet was suspended in 1 ml hypotonic solution (0.8 mM $MgCl_2$, 1.0 mM $CaCl_2$ 30 mM glycerol) and left for 4 min at room temperature. After centrifugation the hypotonic solution was discarded except for one droplet in which the cells were gently resuspended before fixation for 5 min in methanol:acetic acid (3:1). The cells were then dropped on clean glass slides and air-dried. The slides were stained for 7 min in 5% Giemsa diluted in Sörensens buffer (pH 6.8).

The frequency of micronuclei was estimated by scoring at least 2000 nuclei per dose.

RESULTS

Binding of ^{125}I-T3 in GC and CHO cells

The concentration-dependent binding of ^{125}I-T3 in GC and CHO cells is illustrated in Figures 2 and 3. At increasing radioactive concentrations the cellular uptake (cpm per cell) was continuously increased for both types of cells. Contrariwise, nuclear uptake (cpm per nucleus) reached a plateau level after a concentration of about 7.4×10^4 Bq per ml where approximately 30% of the total uptake in GC cells was found in the nucleus. The corresponding value for CHO cells was 2%. The cellular uptake of ^{125}I-T3 seemed to be slightly higher in CHO cells. However, more important is the nuclear uptake which is more than 10 times higher in GC cells. This result reflects the differences between GC and CHO cells in the number of nuclear receptors per cell.

Accumulation of DNA strand breaks

These studies were mainly performed with GH1 cells which have the same origin as GC cells. The cell cycle for GC cells is 24 hours while it is much longer for GH1 cells. The cells, incubated in ^{125}I-T3 containing medium, showed an accumulation of DNA strand breaks Figure 4. At 0.1 cpm per cell about 0.085 breaks per 10^9 daltons were left nonrejoined. The number of ^{125}I-decays in the nucleus was estimated to be just below 0.1 dpm or about 150 decays per 24 hours. Accordingly between 2-3 breaks per ^{125}I-decay appear to be left nonrejoined (assuming 6×10^{-12} g DNA per cell).

Induction of micronuclei

The frequency of micronuclei in GC cells treated at three different radioactive concentrations of ^{125}I-T3 is illustrated in Figure 5. The cells were incubated in radioactive medium for various periods. The number of micronuclei increased with enhanced radioactive concentrations. At 3.7×10^4 Bq per ml about 20 micronuclei were found per 1000 cells (control cells withdrawn). This radioactive concentration would correspond to a nuclear uptake of approximately 0.02 cpm (Figure 2) which is too low to induce any high frequency of micronuclei. Irradiation with 2 Gy of gamma radiation produced about 80 micronuclei per 1000 cells (data not shown).

DISCUSSION

These studies indicate that ^{125}I-decays induce serious DNA damage in GH1 or GC cells after incubation with ^{125}I-T3. The number of nonrejoined DNA breaks per ^{125}I-decay is surprisingly high compared to the effects of ^{125}IdU in DNA (Sundell-Bergman & Johanson, 1980). This might indicate that ^{125}I is located in the close proximity to DNA when ^{125}I-T3 is bound to the receptor. The nonrejoined DNA breaks are probably double strand breaks or even more severe damage to DNA. Single strand breaks are supposed to be repaired continuously during the incubation at 37^0C, since the dose-rate was very low.

We have previously shown that also in various Chinese hamster cell lines decays of ^{125}I-T3 induced nonrejoined DNA strand breaks (Sundell-Bergman & Johanson, 1982). The scattering in these results were higher than in the experiments with GH1 cells probably depending on the experimental conditions such as the use of serum depleted medium. Nevertheless ^{125}I-T3 seemed to be as efficient as $^{125}IUdR$ in inducing serious DNA damages even in Chinese hamster cells.

The time-dependent uptake of ^{125}I-T3 in cells is rather fast reaching a plateau after 60 minutes (Sundell-Bergman & Johanson, 1983). Therefore the decay rate of ^{125}I in the cells can be assumed to be uniform during the incubation. Increasing radioactive concentrations increased the cellular uptake while the nuclear uptake was saturated after about 7.4×10^4 per ml. It appears likely that, in the nucleus, ^{125}I-T3 is only taken up by the T3 receptor and subsequently bound to the chromatin. The nonspecific binding of ^{125}I-T3 in the nucleus is less than 2% (Mitsuhashi et al., 1987).

Whether the total cellular uptake is specific or not remains unclear. Cytoplasmic receptors have been reported but no saturable binding with increasing concentrations of ^{125}I-T3, were found in our studies (Horiuchi et al., 1982). This might be interpreted as indicating the existence of non-specific cytoplasmic binding.

Micronuclei are usually scored after one cell cycle has elapsed when the peak frequency occurs (Midander & Revesz, 1980). Since radiation also induces mitotic delay the appearance of micronuclei may be further delayed. After external gamma irradiation (high dose-rate) of GC cells the peak was observed 28 hr postirradiation. Using internal irradiation at low dose rate it is much more difficult to establish the time for the maximum response. Hence, the frequency of micronuclei were determined at three different sampling times following labelling but only minor differences could be observed (data not shown).

Thyroid hormones induce a wide variety of metabolic effects which have been studied in great detail in liver cells (Bernal et al., 1978). The hormone is also very important for differentiation processes e.g., metamorphosis in amphibians (Robinson, 1977) and for the induction of growth hormone production (Samuels et al., 1976). A possible role of T3 in brain development has been suggested and the presence of nuclear T3 receptors in human fetuses at mid-gestation support this assumption (Bernal & Pekonen,

1984). In fact the concentration of nuclear receptors in the brain was found to increase 10 times from the 10th to the 16th - 18th week. This phase of brain growth is very critical and the cells are extremely radiosensitive. However, the uptake of radioactive labelled T3 is very low (Johanson, unpublished results) and concomitantly, the effects of ^{125}I-T3 might be difficult to detect. Still the combination of T3 responsive cells in high radiosensitive organs might have an impact on the radiation risks estimation for some Auger emitting iodine isotopes.

ACKNOWLEDGEMENTS

Supported by grants from the Swedish Natural Science Research Council. We thank Mrs. U. Johanson and Mr. B. Sandström for excellent technical assistance.

Figure 1. A model of the binding of thyroid hormones (T3/T4) in mammalian cells. The hormone binding region is supposed to be localized near the carboxyl terminal end of the receptor molecule, while the putative DNA binding region is to be found in the cystein rich sequences.

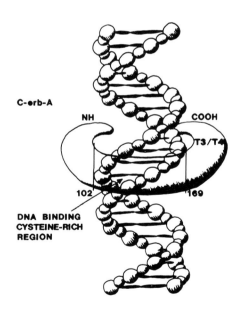

Figure 2. Cellular respectively nuclear uptake of ^{125}I-T3 in GC cells. Cells were incubated with various radioactive concentrations of ^{125}I-T3 for 24 hr.

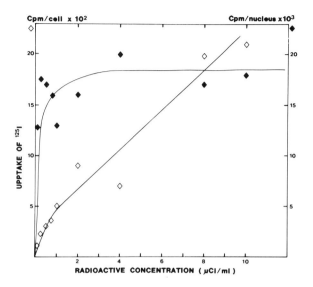

Figure 3. Cellular respectively nuclear uptake in CHO cells. Cells were incubated with various radioactive concentrations of ^{125}I-T3 for 24 hr.

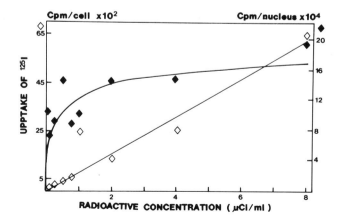

Effects of ^{125}I-labelled thyroid hormones 155

Figure 4. Accumulation of nonrejoined DNA strand breaks in GH1 cells after 24 hr of incubation with various radioactive concentrations of ^{125}I-T3 (from Sundell-Bergman & Johanson, 1983).

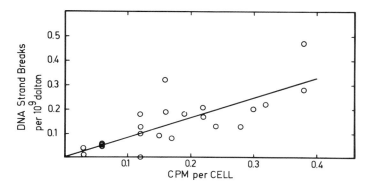

Figure 5. Frequency of micronuclei per 100 GC cells. The cells were incubated with three different radioactive concentrations of ^{125}I-T3 for 72, 96 or 144 hr. Sampling time was 27 hr post labelling. Each curve represents one single experiment.

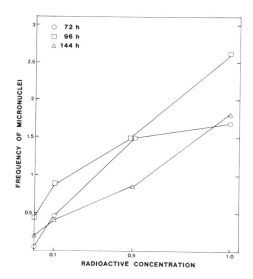

REFERENCES

Apilletti, J.W., David-Inouye, Y., Baxter, J.D. & Eberhardt, N.L., 1983, In Molecular Basis of Thyroid Hormone Action, edited by J.H. Oppenheimer & H.H. Samuels (Academic Press), p.67.

Bernal, J., Refetoff, S. & DeGroot, L., 1978, Journal of Clinical Endocrinology and Metabolism, 47, 1266.

Bernal, J. & Pekonen, F., 1984, Endocrinology, 114, 677.

Bradley, E.W., Chan, P.C. & Adelstein, S.J., 1975, Radiation Research, 64, 555.

Burki, H.J., Roots, R., Feinendegen, L.E. & Bond, V.P., 1973, International Journal of Radiation Biology, 24, 363.

Chan, P.C., Lisco, E., Lisco, H. & Adelstein, S.J., 1976, Radiation Research, 67, 332.

Charlton, D.E., 1986, Radiation Research, 107, 163.

Crantz, F.R., Silva, J.E. & Larsen, P.R., 1982, Endocrinology, 110, 367.

Gibbs, R., Camakaris, J., Hodgson, G. & Martin, R., 1987, International Journal of Radiation Biology, 51, 193.

Horiuchi, R., Johnson, M., Willingham, M., Pastan, I. and Cheng, S-Y., 1982, Proceedings of the National Academy of Science (USA), 79, 5527.

Liber, H.L., LeMotte, P.K. & Little, J.B., 1983, Mutation Research, 111, 387.

Midander, J. & Revesz, L., 1980, International Journal of Radiation Biology, 38, 237.

Mitsuhashi, T., Uchimura, H. & Takaku, F., 1987, The Journal of Biological Chemistry, 262, 3993.

Oppenheimer, J.H., 1979, Science, 203, 971.

Oppenheimer, J.H., Schwartz, H.L. & Surks, M.I., 1973, Biochemical and Biophysical Research Communications, 55, 544.

Oppenheimer, J.H., Schwartz, H.L. & Surks, M.I., 1974, Endocrinology, 95, 897.

Robinson, H., 1977, In Thyroid Hormones and Brain Development, edited by G.D. Grave (Raven Press), P.137.

Rydberg, B., 1975, Radiation Research, 61, 274.

Samuels, H.H., 1983, In Molecular Basis of Thyroid Hormone Action, edited by J.H. Oppenheimer & H.H. Samuels (Academic Press), p. 36.

Samuels, H.H., Stanley, F. & Shapiro, L., 1976, Proceedings of the National Academy of Science USA, 73, 3877.

Sap, J., Munoz, A., Damm, K., Goldberg, Y., Ghysdael, J., Leutz, A., Beug, H. & Vennström, B., 1986, Nature, 324, 635.

Schmid, W., 1975, Mutation Research, 31, 9.

Sundell-Bergman, S., Bergman, R. & Johanson, K.J., 1985,

Mutation Research, 149, 257.
Sundell-Bergman, S. & Johanson, K.J., 1980, Radiation Environmental Biophysics, 18, 239.
Sundell-Bergman, S. & Johanson, K.J., 1982, Biochemical and Biophysical Research Communications, 106, 546.
Sundell-Bergman, S. & Johanson, K.J., 1983, In Proceedings of Biological Effects of Low-level Radiation, IAEA-SM-266/50 (Vienna), p. 191.
Weinberger, C., Thompson, C.C., Ong, E.S., Lebo, R., Gruol, D.J. & Evans, R.M., 1986, Nature, 324, 641.
Weirich, R., Schwartz, H. & Oppenheimer, J., 1987, Endocrinology, 120, 664.
Yaffe, B. & Samuels, H., 1984, The Journal of Biological Chemistry, 259, 6284.

GENE AMPLIFICATION IN CHINESE HAMSTER EMBRYO CELLS
BY THE DECAY OF INCORPORATED IODINE-125

Christine Lücke-Huhle, Angelika Ehrfeld and Waltraud Rau

Kernforschungszentrum Karlsruhe,
Institut für Genetik und Toxikologie,
Postfach 3640, D-7500 Karlsruhe 1, FRG.

ABSTRACT

Simian Virus 40-transformed Chinese hamster embryo cells (Co631) contain 5 viral copies integrated per cell genome. These SV40 sequences were used as an endogenous indicator gene to study the response of mammalian cells to radiation at the gene level. Cells were internally irradiated by Auger electrons emitted by Iodine-125 which was incorporated in the cell DNA in form of 5-$[^{125}I]$ iododeoxyuridine (^{125}IdU). An increase in gene copy number was measured using dispersed cell blotting and Southern analysis in combination with highly sensitive DNA hybridization. A 13-fold amplification of the SV40 sequences and a 2-fold amplification of two cellular oncogenes of the ras family were found. Other cellular genes, like the α-actin gene, are not amplified and no variation in gene copy number has been observed after incubation of cells with cold IdU. Thus, specific gene amplification seems to be the consequence of radiation-induced DNA damage and the resulting cell cycle arrest.

INTRODUCTION

The radiation toxicity of incorporated Iodine-125 is caused mainly by the emitted Auger electrons which are characterized by high LET and small penetration depth. Therefore, most of the dose is absorbed by the DNA itself and high RBE values have been described for endpoints relevant to carcinogenesis, such as chromosomal aberrations, mutations and cell transformation (review: Pomplun et al., 1986). We are interested in one of the early effects of Iodine-125 disintegration on genes: the radiation-induced selective increase in gene copy number.

Normally, all mammalian genes are replicated once during

the S-phase of the cell cycle and two genes of each kind
(one from the father and one from the mother) are passed on
to each daughter cell. However, quantitative changes in
gene copy number are frequently found in biology. They
range from polyploidization of chromosomes (Nagel, 1978) to
selective amplification of single genes (Schimke, 1984). In
some cases gene amplification is associated with
chromosomal abnormalities such as double minute chromosomes
(DM's) or homogeneously staining regions (HSR's). Both
karyotypic alterations are frequently found in tumour cells
(Balaban-Malenbaum & Gilbert, 1977).

Current interest in the phenomenon of gene amplification
stems from finding amplified oncogenes in tumour cells
(Schwab et al., 1983), the fact that gene amplification has
been shown to be responsible for oncogene activation (Marx,
1984) and that DNA damaging agents, like most of the known
carcinogens can induce gene amplification (Lavi, 1981;
Schlehofer et al., 1983; Lücke-Huhle et al., 1986).
Although a direct connection has yet to be established
between quantitative changes in DNA as the initial event
and later tumour development, the possible involvement of
gene amplification in radiation carcinogenesis is an
exciting new development.

External radiation like 60-Co-Gamma rays, 241-Am-Alpha
particles or UV light induces amplification of the viral
DNA sequences in SV40-transformed Chinese hamster embryo
cells with an efficiency independent of the kind of DNA
damage (Lücke-Huhle et al., 1986). Here we show that
radiation originating from the decay of Iodine-125
incorporated into the DNA induces amplification of the SV40
sequences and of two oncogenes of the ras family (Kirsten
and Harvey-ras). The two oncogenes were chosen because it
has been shown that an increase in ras gene product has a
transforming effect on cells (DeFeo et al., 1981).

MATERIAL AND METHODS

Cell culture. SV40-transformed Chinese hamster embryo
cells (Co631; courtesy S. Lavi, Tel Aviv) containing 5 SV40
copies per cell genome (Lavi, 1981) were grown as
monolayers in Eagle's MEM supplemented with 10% fetal calf
serum and 100 U/ml penicillin and 100 µg/ml streptomycin.

Labelling procedure. 5-[^{125}I] Iodo 2-deoxyuridine
(^{125}IdU) with a specific activity of 8.14×10^7 MBq/mmol
was purchased from New England Nuclear (NEN, Boston).
Exponentially growing cells were incubated for 24 hr with
^{125}IdU, the activity ranging from 92.5 to 1850 Bq/ml. Since
the generation time of Co631 cells is 20 hr, in most of the

cells thymidine was partly replaced by ^{125}IdU during a 24 hr incubation. Subsequently monolayers were washed and further incubated in fresh medium or directly assayed for cell survival by their colony forming ability (Ehrfeld et al., 1986).

Gene amplification assay. The amplification of specific genes was analyzed by the dispersed cell assay (Lavi & Etkin, 1981). Cell monolayers were trypsinized at various times (1-7d) after ^{125}IdU incorporation and samples of 5×10^5 cells were trapped on 25 mm diameter nitrocellulose filters. Cells were lysed and their DNA denatured by soaking the filters three times in 0.5 M NaOH/1M NaCl and then neutralizing with 0.5 M Tris HCl/3 M NaCl, pH 7.4. Filters were then dried at $80°C$ in vacuo and hybridized with ^{32}P-labelled DNA probes.

Southern transfer hybridization. For these experiments high molecular weight DNA was isolated from ^{125}IdU treated and control Co631 cells and subjected to electrophoresis on a 0.8% agarose gel. By this method DNA fragments of different size can be separated with high resolution. After denaturation in 1 N NaOH these fragments were transferred from the gel to a nitrocellulose filter by the Southern transfer technique (Southern, 1975) described in detail by Lücke-Huhle et al., (1986).

Hybridization procedure. Filters were preincubated at $42°C$ in a solution containing 50% deionized formamide, 0.05 M Pipes buffer, pH 6.5, 5 x Denhardt's solution (1 x Denhardt's solution contains 0.02% bovine serum albumin plus 0.02% polyvinyl-pyrrolidone plus 0.02% Ficoll 400). 0.5 mg/ml yeast t-RNA and 5x SSC (1x SSC is 0.15 M NaCl/0.01 M tri-Na-citrate) for 15 hr and afterwards hybridized with ^{32}P-DNA probes labelled in vitro by nick-translation (Rigby et al., 1977) with ^{32}P-CTP to 10^8 cpm/µg DNA. The DNA probes used were a) SV40-DNA (BRL, Cat No 5252 SA/SB), b) α-actin plasmid DNA (Minty et al., 1981), and two cellular oncogenes of the ras-group: c) Kirsten-ras cloned in pBR322 (Ellis et al., 1981) and d) a clone of the activated Harvey-ras gene from a human bladder carcinoma line, also cloned in pBR322 (Parada et al., 1982; Tabin et al., 1982). Autoradiograms were made with Kodak X-Omat AR films. For quantitation of the β-radioactivity filters were analysed in a liquid scintillation counter (Packard Instruments, Tri Carb 2660).

RESULTS AND DISCUSSION
Following incorporation of ^{125}IdU, SV40-transformed

Chinese hamster embryo cells show a selective amplification of the integrated SV40 sequences without producing intact virus. As determined by the dispersed cell assay the number of SV40 copies per cell genome increases with increasing dose (Figure 1). The signal seen at dose 0 reflecting the 5 SV40 copies integrated per cell genome increases with increasing dose. This test is based on the fact that denatured cellular DNA binds copies of a specific ^{32}P-labelled gene probe in proportion to the number of copies of the same sequence that is present in the genome.

Since the 35 keV γ-rays emitted by the ^{125}I isotope in the higher dose range also blackens the sensitive Kodak film, an unhybridized half of each filter was placed underneath the one hybridized with the ^{32}P-labelled DNA-probe. For quantitation of the bound radioactivity both filter halves were measured in a β-scintillation counter using the discriminator setting for ^{32}P. Under these conditions the counts for the unhybridized filters were below background level.

Amplification of SV40 sequences also increases with time after ^{125}IdU labelling (Table 1). For the higher doses an increase in SV40 copy number is detectable at day 3 after labelling and increases steadily until day 7, the longest time interval investigated in this study.

Amplification factors were calculated in comparison to the signal given by the untreated control cells. For SV40 DNA the factors ranged from 3.4 to 12.9 at day 7 after 2.5 and 1850 Bq/ml, respectively.

Absolute dose calculations in terms of decays per cell were difficult because of the different numbers of cell divisions occurring during the time of the experiment in cultures exposed to various doses of ^{125}IdU. The ^{125}IdU activities ranging from 92.5 to 1850 Bq/ml which were used in the 24 h incubation, and the corresponding survival ratios (N/No) are given in Figure 1. Survival decreases exponentially with increasing amounts of incorporated ^{125}I (not shown). The D_{37} amounts to 95 decays per cell corresponding to 0.66 Gy (Ehrfeld et al., 1986).

Hybridization of filters from the same set of experiment with ^{32}P-labelled oncogenes, Ki-<u>ras</u> and Ha-<u>ras</u>, shows a smaller but significant increase in the number of gene copies of these cellular homologues by factors of 2.3 and 1.8, respectively (Figure 1, Table 2).

Other cellular genes like the α-actin gene were not amplified (Figure 1) and no variation in gene copy number was detected after incubation of cells with cold IdU in

Figure 1. Autoradiograms showing an increase in gene copy number for SV40 sequences and the two oncogenes Ki-<u>ras</u> and Ha-<u>ras</u> with increasing dose of ^{125}I (see Ehrfeld et al., 1986). The blackening of the film in the upper halves of the filters is caused by the ^{32}P-labelled gene probes hybridized to homologous cellular DNA sequences in the Chinese hamster genome, the lower halves are unhybridized and their uniform blackening, seen mainly at higher doses, is due to ^{125}I-disintegrations themselves. ^{125}IdU is still present in the DNA on the filters. In each case the blackening of the lower half of the filter must be subtracted from that of the upper half to obtain the amplification effect.

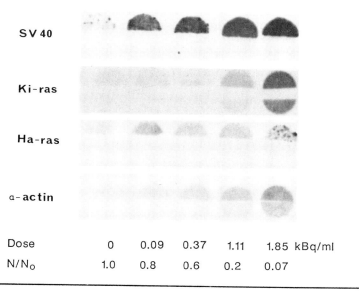

equimolar concentrations (not shown).

Southern analysis of undigested genomic DNA isolated 3 days after labelling with 1850 Bq/ml ^{125}IdU supports the cell blot data. It also shows amplification of the integrated SV40 sequences in the high molecular weight fraction (>70 kb) and with multiple heterogeneous DNA fragments (Figure 2), the latter showing up mainly as a smear in the autoradiogram (lane 3). In addition, the existence of defined bands of unit size SV40 DNA (form II and form I SV40 molecules of 5.6 and 2.7 kb in size, respectively) and smaller SV40 containing molecules of 5.0 and 2.2 kb in size, indicates that part of the amplified SV40 DNA has been cut out from the hamster genome. Thus,

Table 1. Amplification factors for SV40 DNA after different doses of ^{125}IdU and various times after treatment.

^{125}IdU kBq/ml	Days after ^{125}IdU labelling							
	0	1	2	3	4	5	6	7
0.0	1.0	1.0	1.0	1.0	1.0	1.0	1.0	1.0
0.093	n.d	1.1	0.8	n.d	1.1	1.4	1.5	3.4
0.37	0.6	1.0	0.8	1.0	1.3	2.1	1.9	5.2
1.11	0.7	0.8	1.0	1.7	2.7	5.1	3.4	7.5
1.85	1.4	1.1	n.d	3.2	4.2	5.6	7.8	12.9

Note Incubation of cells for 24 hr in .09, .37, 1.11 and 1.85 kBq ^{125}IdU/ml medium led to an uptake of .15, .6, 1.8 and 3.0 mBq/cell, respectively. The gene amplification factors are calculated by comparing the gene copy number of treated cells to control cultures (n.d. = not determined).

Table 2. Amplification factors for various cellular genes following ^{125}IdU incubation and 72 hr post incubation in fresh medium.

^{125}IdU kBq/ml	Amplification factors		
	Ki-<u>ras</u>	Ha-<u>ras</u>	α-actin
0.0	1.0	1.0	1.0
0.093	1.6	1.5	0.8
0.37	1.4	1.5	1.0
1.11	1.9	1.6	1.0
1.85	2.3	1.8	0.9

excision of SV40 molecules occurred although the DNA was incompetent for complete virus formation.

The radiation-induced gene amplifications described here are in accordance with amplifications obtained after external radiation (Lücke-Huhle et al., 1986), after treatment with chemical carcinogens (Lavi, 1981) and after viral co-transfection (Schlehofer et al., 1983), stressing the point that various kinds of DNA damage can induce this effect. In the attempt to elucidate the mechanism of gene amplification, the role of the origin of replication has been well established in the case of cell lines transformed by SV40 (Lavi, 1981, Lücke-Huhle and Herrlich, 1986) and one might suggest that blockage of cellular DNA replication causes repeated reinitiation at specific origins within one

Figure 2. Autoradiogram of Southern blots from isolated DNA (10 μg each) of control cells (lane 2) and irradiated cells (lane 3) 3 days after labelling with 1850 Bq/ml ^{125}IdU). In lane 1 SV40 marker DNA gives the position of SV40 form I and form II DNA at 2.7 and 5.6 kb, respectively.

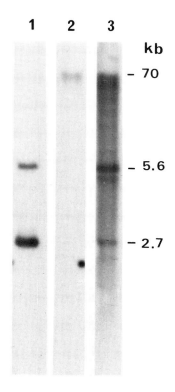

cell cycle (Schimke, 1984).

Malignant transformation is a complex process. It is generally agreed that several steps are involved: an initiation, promotion, and progression step. According to current ideas gene amplification could play a role in all three. Amplification of oncogenes could be imagined as an initial event; genomic rearrangements subsequent to amplification of any gene might promote transformation by changing the expression of critical genes in the new neighbourhood, and, from tumour data it is known that the degree of amplification of oncogenes goes along with the severity of the disease (Schwab, 1986).

ACKNOWLEDGEMENT

The authors wish to thank Professor D.M. Taylor for critically reading the manuscript and Miss Monika Pech for excellent technical assistance.

REFERENCES

Balaban-Malenbaum, G. & Gilbert, F., 1977, Science, 198, 739.

DeFeo, D., Gonda, M.A., Young, H.A., Chang, E.H., Lowy, D.R., Scolnick, E.M. & Ellis, R.W., 1981, Proceedings of the National Academy of Science USA, 78, 3328.

Ehrfeld, A., Planas-Bohne, F. & Lücke-Huhle, C., 1986, Radiation Research, 108, 43.

Ellis, R.W., Defeo, D., Shih, T.Y., Gonda, M.A., Young, H.A., Tsuchida, N., Lowry, D.R. & Scolnick, E.M., 1981, Nature, 292, 506.

Lavi, S., 1981, Proceedings of the National Academy of Science USA, 78, 6144.

Lavi, S. & Etkin, S., 1081, Carcinogenesis, 2, 417.

Lücke-Huhle, C. & Herrlich, P., 1986, Gene amplification in mammalian cells after exposure to ionizing radiation and UV. In: F.J. Burns, A.C. Upton and G. Silini (eds.), Radiation carcinogenesis and DNA alterations, Plenum Press, Amsterdam, p. 405.

Lücke-Huhle, C., Pech, M. & Herrlich, P., 1986, Radiation Research, 106, 345.

Marx, J.L., 1984, Science, 223, 40.

Minty, A.J., Caravatti, M., Robert, B., Cohen, A., Daubas, P., Weydert, A., Gros, F. & Buckingham, M.E., 1981, Journal of Biological Chemistry, 256, 1008.

Nagel, W., 1978, Endopolyploidy and Polyteny in Differentiation and Evolution. North-Holland, Amsterdam.

Parada, L.F., Tabin, C.J., Shih, C. & Weinberg, R.A., 1982, Nature, 297, 474.

Pomplun, E., Booz, J., Dydejczyk, A. & Feinendegen, L.E., 1986. 2nd Symposium on "Molekulare und Zelluläre Mechanismen bei Wirkung von Strahlen", Jülich.

Rigby, P.W.J., Dieckmann, M., Rhodes, C. & Berg, P., 1977, Journal of Molecular Biology, 113, 237.

Schimke, R.T., 1984, Cell, 37, 705.

Schlehofer, J.R., Gissmann, L., Matz, B. & zur Hausen, H., 1983, International Journal of Cancer, 32, 99.

Schwab, M., Alitalo, K., Varmus, H.E., Bishop, J.M. & George, D., 1983, Nature, 303, 497.

Schwab, M., 1986, Amplification of proto-oncogenes and its possible role in tumour progression. In: Proceedings of the workshop on: The role of DNA amplification in tumor initiation and promotion. Heidelberg, in press by Lippincott.

Southern, E.M., 1975, *Journal of Molecular Biology*, 98, 503.

Tabin, C.J., Bradley, S.M., Bergman, C.J., Weinberg, R.A., Pagageorge, A.G., Scolnick, E.M., Dhar, R., Lowy, D.R. & Chang, E.H., 1982, *Nature*, 300, 143.

LETHAL, MUTAGENIC AND DNA-BREAKING EFFECTS OF DECAYS OF
IODINE-125 UNIFILARLY INCORPORATED INTO THE DNA OF
MAMMALIAN CELLS[1]

Yoshisada Fujiwara and Noriyuki Miyazaki[2]

Department of Radiation Biophysics,
Kobe University, School of Medicine,
Kusunoki-cho 7-5-1, Chuo-ku, Kobe 650, Japan

ABSTRACT

Decays of [125-I]iodo-2'-deoxyuridine [125-I]dU incorporated unifilarly into the DNA of S-phase in synchronized Chinese hamster cells were very effective for cell killing and mutagenesis as characterized by the RBE values of 14 and 20, respectively. Each decay of 125-I in the unifilarly labelled DNA while kept frozen caused about 4 single-strand breaks (ssbs) and nearly 1 double-strand break (dsb) at the site of decay, the multi-break production. About a half of such DNA breaks were less easily rejoined within 6 hr. Thus, the efficient production of a dsb by the high local energy deposition per decay is the basis for the extraordinarily high lethal and mutagenic effects.

INTRODUCTION

Iodine-125 disintegrates by electron capture and internal conversion with the emission of 13-21 Auger electrons (0.7-70 keV, mean ~19 keV) (Charlton & Booz, 1981). In the DNA duplex-simulating cylindrical target, 125-I decay deposits locally a greater mean energy, 253 eV, than that (mean 65 eV) from 5 MeV alpha-particles (Charlton, 1986). Therefore, Auger electrons will exhibit a high LET-like effect on the DNA in the immediate vicinity of the decay. In fact, 125-I decay causes strand breaks mainly between 1 and 5 nucleotides in both strands from the

---

1. Supported in part by a Grant-in-Aid from the Ministry of Education, Science and Culture, Japan.
2. Present address: RI Division, Psychiatric Research Institute of Tokyo, Kamikitazawa 2-1-8, Setagaya-ku, Tokyo 156.

decay in purified DNA (Martin & Haseltine, 1981) and marked lethality, ssbs and dsbs in coliphage, E. coli, and mammalian cells (Burki et al., 1973, Hofer et al., 1977; Krisch et al., 1976; Liber et al., 1983; Miyazaki & Fujiwara, 1981; Schimid & Hotz, 1973).

Mutagenic efficiency of [125-I]dU incorporated into DNA as a function of survival is lower than that of [3-H] decays and external gamma-rays in E. coli (Ahnström et al., 1970). In contrast, 125-I decay in the DNA causes markedly enhanced hprt mutation in mammalian cells, compared to X(gamma)-rays (Miyazaki & Fujiwara, 1981; Liber et al., 1983). Further, mutagenic efficiency of 125-I as a function of survival is higher than that by [3-H] decays and X-rays in human cells (Liber et al., 1983), suggesting that the mutagenic efficiency of DNA damage by different radiations differs.

In this V79-cell study, we attempt to determine the lethal and mutagenic efficiencies of 125-I decays and the survival-mutation relationship under the refined condition of complete unifilar labelling of DNA with [125-I]dU, and further to delineate the relation between the decay-induced dsb and such biological endpoints.

MATERIALS AND METHODS
Cell line and synchronization

V79 Chinese hamster clone 743-367 was grown in Dulbecco's modified minimal essential medium (DMEM) with 10% fetal bovine serum (FBS) (Fujiwara & Tatsumi, 1980). Since the doubling time was about 12 hr, cells were synchronized at G1/S by treating with 1 µM 5-fluoro-2'-deoxyuridine (FdU) plus 5 µM deoxycytidine (dC) for approximately one generation time of 14 hr. The S-phase length after release of FdU block was 8.0 hr.

Unifilar labelling of DNA with [125-I]dU or IdU

The cells synchronized at G1/S (see above) were released to synthesize DNA only during S phase by incubating for 10 hr in 10 µM [125-I]dU, 10 µM IdU, or 10 µM thymidine (dT) in the continuous presence of FdU-plus-dC. For unifilar labelling for cell killing and mutation, 10 µM [125-I]dU of a low specific activity contained 370 Bq/ml of [125-I]dU (1.74×10^5 MBq/mg) and 10 µM nonradioactive IdU. The amount of incorporation was 3.7×10^{-5} Bq/cell, or roughly 5 decays per cell (dpc)/day. For detecting 125-I decay-induced ssbs, cells were first prelabelled with [14-C]dT (925 Bq/ml, 1.57×10^5 MBq/mmole) for 2 days, chased for 2 hr, synchronized at G1/S, and then incubated in 10 µM [125-I]dU of a high specific activity (0.611 MBq/ml, 1.67×10^5 MBq/mg) in the presence of FdU-plus-dC for a longer period of 14 hr due to delayed S-phase traverse.

The DNA extracted from such labelled cells was assessed for the degree of IdU replacement by neutral and alkaline CsCl density centrifugation (Fujiwara, 1974). The analysis revealed the 95% of DNA banded at the IdU/dT hybrid density in neutral CsCl and at least 95% [125-I]dU replacement in heavy daughter strands, as estimated from the density increase between the light parental [14-C]dT strands and the heavy new [125-I]dU strands in alkaline CsCl. Thus, the above labelling protocol by use of synchronized G1/S cells and the FdU block to endogenous dTMP synthesis ensured almost complete unifilar labelling, which was better than the previous 80-90% replacement by a 14 hr labelling of exponentially growing cells (Miyazaki & Fujiwara, 1981).

Determination of 125-I decays

After labelling, cells were kept frozen in the medium (DMEM, 10% FBS, 10% DMSO) at a final density of 10^6 cells per vial at -79^0C to accumulate 125-I decays for desired lengths of time. To determine 125-I incorporation into DNA, triplicate samples of 10^5 cells were washed with 5% cold trichloroacetic acid, and acid-insoluble materials were assayed for 125-I and 14-C radioactivity by the dual counting program in a Beckman liquid scintillation spectrometer. The initial uptake (Bq/cell) and the numbers of dpc were determined. 125-I decays (d) were converted to absorbed dose (D) by D = 0.013d (Gy) for a frozen V79 cell nucleus (Burki et al., 1973).

External X-irradiation

The unifilarly labelled IdU/dT cells or the control dT/dT cells were irradiated with 170 kVp X-rays (dose rate: 1 or 12 Gy) while frozen in dry ice/ethanol at about -70^0C.

Survival and mutation assays

Both optimum assay methods have been described in detail elsewhere (Miyazaki & Fujiwara, 1981). Survival curves were characterized by extrapolation number (n) and mean lethal dose (Do). 6-Thioguanine-resistant (hprt gene) mutations induced by 125-I decays in the unifilarly labelled [125-I]dU/dT-DNA were assayed after the maximum phenotype expression period of nine days and compared with those in the dT/dT and IdU/dT cells by external X-irradiation at -70^0C.

Alkaline sucrose sedimentation of DNA

V79 cells with accumulated 125-I decays (7.46×10^4 dpc) in the [125-I]dU/[14-C]dT-DNA at -79^0C were lysed immediately after thawing, or further incubated for 3 and 6 hr at 37^0C before cell lysis. After exhaustive removal of dead cells and cell debris, more than 95% of live cells were lysed in 0.25% sodium dodecyl sulphate, 10 mM EDTA and 0.15 M sodium bicarbonate (pH 8.0) and digested with 2 mg/ml Pronase for 4 hr at 37^0C. A 0.2-ml aliquot of the finally alkalinized lysate (7,000 cells/0.2 ml) was layered on top of 4.8 ml of a 5-20% (w/v) alkaline sucrose gradient and centrifuged at 35,000 rpm for 1 hr in an SW50.1 rotor of a Beckman Model L5-50 ultracentrifuge, followed by isolation of 25 fractions for assay of radioactivity of the dual [14-C]/[125-I] labels in acid-insoluble DNA (Fujiwara, 1987). The weight average mol. wt. (Mw) was first computed from sedimentation profile (see Fig. 4) and then the number average mol. wt. (Mn) was calculated as Mn = 0.5 Mw. The numbers of ssb per dalton and per V79 cell (N) were calculated respectively as

N (ssbs/dalton) = [1/Mn(o)]   (1)
N (ssbs/cell) = [1/Mn(t) - 1/Mn(o)] x (3×10^{12})   (2)

where Mn(o), Mn(t) and the constant 3×10^{12} represent Mn of control DNA, Mn of treated DNA, and DNA content in dalton per V79 cell, respectively.

RESULTS

Effect of Auger electrons quantified by cell survival

We obtained the uniform unifilar [125-I]dU incorporation and a low value of ~5 dpc per day while frozen by use of a low specific activity under the more refined condition than our previous experiments (Miyazaki & Fujiwara, 1981). Figure 1 shows the survival curves after various accumulated decays in the [125-I]dU/dT-DNA at -79^0C and after low-LET X-ray exposure to the dT/dT or IdU/dT cells at -70^0C. In the cells with [125-I]dU/dT-DNA, the survival curve of the least squares fit after 125-I decays was an exponential high-LET type without the threshold shoulder, characterized by n = 1.0 and Do = 38.3 ± 1.3 dpc or 0.50 Gy (Figure 1). The Do value is slightly smaller than our previous one (= 40 dpc) from the curve with large scatter of data points (Miyazaki & Fujiwara, 1981).

The external X-ray survival curves revealed larger Do values such as n = 2.2, Do = 7.0 Gy for the dT/dT cells and n = 1.0, Do = 5.0 Gy for the IdU/dT cells at -70^0C (Figure 1). We have already determined that the dT/dT and IdU/dT

Figure 1. Survival curves of V79 cells after 125-I decays in [125-I]dU/dT-DNA at -79^0C and after X-irradiation of IdU/dT and dT/dT cells at -70^0C.

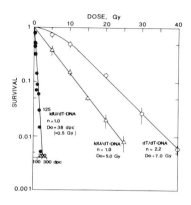

Table 1. Mean lethal dose (Do), 10%-survival dose (D_{10}) and relative biological effectiveness (RBE).

Parameters	125-I decay [125-I]dU/dT	X-rays to IdU/dT	dT/dT
Frozen at -70^0C or -79^0C			
Do (Gy)	0.50	5.00	7.00
RBE	14 (10)[a]	(1)	1
D_{10} (Gy)	1.00	11.80	21.00
RBE	21 (12)	(1)	1
Unfrozen (20^0C)[b]			
Do (Gy)	nt[c]	0.65	1.40
D_{10}	nt	1.30	5.30
Do(-70^0C)/Do(20^0C)		8	5

a: The number in parentheses are RBEs compared with X-rays in IdU/dT cells.

b: From Miyazaki & Fujiwara (1981).

c: Not tested.

V79 cells had n = 5.2, Do = 1.4 Gy and n = 1.0, Do = 0.65 Gy at 20°C room temperature (Miyazaki & Fujiwara, 1981) (see Table 1). The -70°C freezing restricts the mobility of X-ray-induced free radicals and their interactions with molecular oxygen, thus minimizing the indirect effect. As described above, X-irradiation at -70°C, in fact, generated 8 and 5 times larger Do values for the dT/dT and IdU/dT cells than those at 20°C (Table 1). Thus, freezing eliminates almost all of the 80-90% indirect effect produced at 20°C by low-LET X-rays. At -70°C, the IdU/dT cells were 1.4 times more sensitive (= 7 Gy/5 Gy) to X-rays than the dT/dT cells, suggesting a more direct radiolysis of IdU in DNA.

The above Do comparison in the frozen state indicates that RBEs of 125-I decays are 14 and 10, compared with X-rays in dT/dT and IdU/dT cells, respectively, for cell inactivation (Table 1). RBEs based on D_{10} are much larger (Table 1).

Mutagenesis

Figure 2 shows the dose-mutation induction relationship in a range of 95-20% survival. In the frozen dT/dT cells, the X-ray doses up to 10 Gy that covered only the shoulder region (Figure 1) caused the lowest curvilinear induction of hprt mutations (Figure 2). However, 125-I decays in the [125-I]dU/dT-DNA at -79°C and external X-irradiation of the IdU/dT cells at -70°C induced hprt mutations as a linear

Figure 2. The hprt mutation induction by 125-I decay and X-irradiation at frozen state.

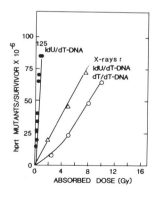

function of dose (Figure 2). In the case of 125-I decays, the regression line indicated the highest mutation induction rate of 11.0 x 10^{-7} mutants/survivor per 0.01 Gy or 14 x 10^{-7}/dpc. On the other hand, the mutation rates by X-rays at $-70^{0}C$ were 1.00 x 10^{-7}/0.01 Gy for the IdU/dT cells and about 0.6 x 10^{-7}/0.01 Gy (a rough estimate) for the dT/dT cells, in agreement with Miyazaki & Fujiwara (1981).

In the frozen state, therefore, RBEs of 125-I decay for mutation are about 20 (= 11.0 x 10^{-7}/0.6 x 10^{-7}) compared to X-rays in the dT/dT cells and 11 (= 11.0 x 10^{-7}/1.00 x 10^{-7}) compared to X-rays in the IdU/dT cells. In addition, X-rays were again twice as mutagenic to the IdU/dT-DNA as they were to the dT/dT-DNA in the frozen state.

Figure 3 shows the induced mutation frequency against the log (survival) or lethal hit in frozen cells. The hprt mutations induced by either 125-I decays or external X-ray exposures to the dT/dT and IdU/dT cells followed the same regression line (r = 0.967). With these agents, one lethal hit to a V79 cell will induce approximately 6 x 10^{-5} hprt mutations. At $20^{0}C$, a similar relationship was also observed after X-irradiation of dT/dT and IdU/dT cells. Thus, DNA lesions by 1 lethal hit will generate gene mutations at a very similar probability, whether X-rays or Auger electrons and direct or indirect effects are involved. Liber et al. (1984) obtained a higher efficiency by 125-I than by X-rays, which is different from Figure 3.

Figure 3. Plots of the induced mutation frequencies against survival or lethal hit. Data from Figures 1 and 2.

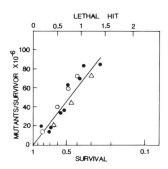

125-I disintegrations and DNA strand breaks

Uniformly [14-C]dT-prelabelled, synchronized V79 cells were labelled with [125-I]dU of a high specific activity. The amount of 125-I incorporated as uniform [125-I]dU/[14-C]dT-DNA was 2.37×10^{-3} Bq/cell, which was allowed to accumulate a total of 7.46×10^4 dpc while kept frozen for 39 days. The cells were then thawed to detect the numbers of decay-induced ssb in both single [125-I]dU and [14-C]dT strands by immediate cell lysis.

Figure 4A shows alkaline sucrose sedimentation profiles of [125-I]dU strands and complementary [14-C]dT strands carrying ssbs induced by Auger electrons from 7.46×10^4 dpc. The [14-C]dT strands of the control IdU/[14-C]dT-DNA had a profile peaking at fraction 10 ($M_w = 2.60 \times 10^8$), as observed in the [14-C]dT/[14-C]dT-DNA (data not shown). The present [14-C]-radioactivity (2,750 cpm/7,000 cells) after the 39-day decay did not cause observable ssbs. After 7.46×10^4 dpc, the sedimentation profiles of both [125-I]dU and [14-C]dT strands shifted to the similar low-molecular-weight position peaking at fraction 21 near top of the gradient (Figure 4A). The [125-I]dU strands had a M_n of 9.50×10^6, being a little smaller than the M_n (= 10.8×10^6) of the complementary [14-C]dT strands (Table 2). Calculation with equation (2) gave 2.93×10^5 and 2.55×10^5 ssbs/cell for the [125-I]dU and [14-C]dT strands, respectively, indicating 3.9 (= $2.93 \times 10^5 / 7.46 \times 10^4$) and 3.4 (= $2.55 \times 10^5 / 7.46 \times 10^4$) ssbs/decay on average for the respective strands in frozen cells. Such figures are

Figure 4. Alkaline sucrose profiles of [14-C]dT strands of control DNA (△), and [125-I]dU (●) and [14-C]dT (○) strands after decays in the [125-I]dU/[14-C]dT-DNA.

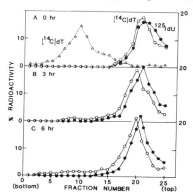

consistent with previous estimates (Painter et al., 1974; LeMotte & Little, 1984). Here, the ratio of ssbs in [14-C]dT strand to those in [125-I]dU strand is 0.87, possibly suggesting the effective induction of dsbs by the cascade of Auger electrons.

Repair of 125-I induced ssbs
Following incubation for 3 and 6 hr at 37^0C, the DNA of the [125-I]dU/[14-C]dT cells was centrifuged, and the results are shown in Figures 4B and 4C, respectively. At 3 and 6 hr, the profile peak of [125-I]dU strands remained unmoved at fraction 21, but it became slightly higher with

Table 2. 125-I decay-induced ssbs and repair.

Incub. time	[125-I]dU strands				[14-C]dT strands			
	Mn	ssb/D	ssb/cell	repair	Mn	ssb/D	ssb/cell	repair
0 hr	9.5	9.76	2.93	0%	10.8	8.99	2.55	0%
3 hr	11.8	7.78	2.31	20%	14.8	5.99	1.80	30%
6 hr	13.5	6.64	1.99	30%	25.0	3.23	0.97	60%

Mn x 10^{-6}; ssb/D (dalton) x 10^{-8}; ssb/cell x 10^{-5}

Table 2 shows that Mn, 9.5 x 10^6 at 0 hr, of the [125-I]dU strands increased slightly to 11.8 x 10^6 at 3 hr and 13.5 x 10^6 at 6 hr. Therefore, there were only small repair fractions of 20% and 30% ssbs at 3 and 6 hr, respectively (Table 2). The complementary [14-C]dT strand exhibited repair of 30% and 60% ssbs at 3 and 6 hr respectively (Table 2). 90-95% of external X-ray induced ssbs in the [14-C]dT strands of the IdU/[14-C]dT-DNA at -70^0C were rejoined much faster, within 1 hr, with a half life of 20 min (not shown). Therefore, clustered multiple ssbs near the 125-I decay site (see above) are less easily repaired. Assuming that ssbs in [14-C]dT strands of the [125-I]dU/[14-C]dT-DNA that remain unrejoined at 6 hr after processing of all repairable ssbs may represent unrejoinable dsbs, the number of dsbs/decay in frozen cells will be 1.3 (= 9.70 x 10^4/7.46 x 10^4). Such a value approaches the reported efficiency of about 1 dsb/decay in both frozen and unfrozen cells (Painter et al., 1974; Krisch et al., 1976; LeMotte & Little, 1984; Charlton & Humm, 1987).

time by the concomitant reduction in [125-I] radioactivity associated with small mol. wt. fractions 22 to 25 (Figures 4B and 4C). This suggests that ssbs in those strands are repaired to some minor extent. In the [14-C]dT strands, the profile shifts at 3 and 6 hr were not so different, and there were 60-70% of the total counts between fractions 15 and 24 (Figures 4B and 4C). We observed a similar sedimentation result after decays for 31 days (not shown).

DISCUSSION

The present results have shown that disintegration of 125-I incorporated into DNA exerts extremely lethal and mutagenic effects in mammalian cells, as judged from the high RBEs, far exceeding those by external high-LET radiation. Such effects arise from less easily repairable dsbs produced by high local energy deposition in the DNA. Charlton (1986) has suggested that 90% of the decays deposit 100 eV or more and 46% deposit more than the average energy of 253 eV. Formation of ssbs peaks at 50-100 eV and instead, further increase in the energy deposition produces dsb and complex breaks (Charlton and Humm, 1987). In support, Figure 4A shows that the ssb frequency in the complementary strands is 87% that in the [125-I]dU strands. Figure 4 and the previous results (Painter et al. 1974; LeMotte & Little, 1984) indicated that about 4 ssbs and nearly 1 dsb are formed by each decay in the DNA of mammalian cells, being almost similar to those in E. coli (Krisch et al., 1976) and phage DNA (Krisch & Sauri, 1975). Painter et al., (1974) observed the repair of only a half of 125-I decay induced ssbs in V79 cells. In accord, about 1/3 and 1/2 of ssbs in the [125-I]dU and complementary strands respectively were repaired within 6 hr, and such repair was very slow (Figure 4). At 6 hr, 60-70% of the total counts in both strands still remained in the main sedimentation profile of unrejoined DNA (Figure 4C). Thus, less easily repairable dsbs are formed between local multiple ssbs on both strands, in agreement with Martin and Haseltine (1981) who showed that more than 70% of the observed strand breaks occurred within 5 nucleotides on both strands from the decay. Such localized strand scissions are induced by Auger electrons, and base lesions result from transmutation (Linz & Stöcklin, 1985).

From our results we can estimate that in the case of 125-I decay, the Do (38.3 dpc = 0.5 Gy) will form 50 dsbs/cell (= 1.3 dsb/decay x 38.3 dpc) and thus 0.01 Gy of Auger electrons generates 1 dsb/cell. On the other hand, we assume that (1) 1 g DNA has 1×10^{21} base pair (bp), (2) 1 cell has about 6×10^9 bp, and (3) the dsb production by

X-rays requires the energy deposition of 600 eV on average. Then, 0.01 Gy of X-rays will produce 0.6 dsb/cell [= (6.24 x 10^{13})(ev/g)/600(eV)/10^{21}(bp) x (6 x 10^9)(bp/cell)]. Thus, the Do (1.4 Gy) at 20^0C (Table 1) will produce the lethal number of 84 dsbs/cell. The Do of X-irradiated frozen cells is 7.0 Gy (Figure 1), and the dsb/Gy efficiency under the frozen condition is 1/5 that in the unfrozen condition (LeMotte & Little, 1984) (Table 1). Therefore, the number of dsbs per Do (= 50-84 dsbs/cell) by X-rays and 125-I decays is not much different whether cells were frozen or not. Highly localised energy deposition by multiple Auger electrons induces 2-3 times more dsbs per decay than sparse energy deposition by X-rays.

Regarding mutation, ionizing radiation is known to induce mainly deletion mutations. In Figure 2, 125-I decay induced an order of magnitude higher hprt mutations on the basis of dose than did external X-rays. Such a high mutagenicity of 125-I decay seems to be related to efficient dsb production. Gibbs et al., (1987) analyzed 125-I decay-induced hprt mutations in CHO cells by Southern blot hybridization and postulated that the predominant initial lesions were minideletions which gave rise to large deletions up to at least 20-30 kbp in most of mutants.

In conclusion, the production of less easily repairable dsb by high local energy deposition by each 125-I decay is the basis of its highly lethal and mutagenic effects.

REFERENCES

Ahnström, G., Ehrenberg, L.E., Hussain, S. & Natarajan, A.T., 1970, Mutation Research, 10, 247.

Burki, H.J., Roots, R., Feinendegen, L.E. & Bond, V.P., 1973, International Journal of Radiation Biology, 24, 363.

Charlton, D.E., 1986, Radiation Research, 107, 163.

Charlton, D.E. & Booz, J., 1981, Radiation Research, 87, 10.

Charlton, D.E. & Humm, J.L., 1987, (in press).

Feinendegen, L.E., Hennerberg, P. & Tisljar-Lentulis, G., 1977, Current Topics of Radiation Research Quarterly, 12, 436.

Fujiwara, Y., 1974, Cancer Research, 32, 2089.

Fujiwara, Y., 1987, Cancer Research, 47, 1118.

Fujiwara, Y. & Tatsumi, M., 1980, Mutation ; Research, 73, 183.

Gibbs, R.A., Camakaris, J., Hodgson, G. & Martin, R.F., 1987, International Journal of Radiation Biology, 51, 193.

Hofer, K.G., Keough, G. & Smith, J.M., 1977, Current Topics

of Radiation Research Quarterly, 12, 335.

Krisch, R.E., Krasin, F. & Sauri, C.J., 1976, International Journal of Radiation Biology, 29, 27.

Krisch, R.E. & Sauri, C.J., 1975, International Journal of Radiation Biology, 27, 553.

LeMotte, P.K. & Little, J.B., 1984, Cancer Research, 44, 1337.

Liber, H.L., LeMotte, P.K. & Little, J.B., 1983, Mutation Research, 111, 387.

Linz, U. & Stöcklin, G., 1985, Radiation Research, 101, 262.

Martin, R.F. & Haseltine, W.A., 1981, Science, 213, 896.

Miyazaki, N. & Fujiwara, Y., 1981, Radiation Research, 88, 456.

Painter, R.B., Young, B.R. & Burki, H.J., 1974, Proceedings of the National Academy of Sciences USA, 71, 4836.

Schmid, A. & Hotz, G., 1973, International Journal of Radiation Biology, 24, 307.

CYTOTOXICITY OF ^{125}I DECAY PRODUCED LESIONS IN CHROMATIN

L.S. Yasui[1], A.S. Paschoa[2], R.L. Warters[1] and K.G. Hofer[3]

[1] Univ. of Utah Health Sciences Ctr., Dept. of Radiology, Salt Lake City, UT 84132.
[2] Pontificia Universidade Catholica, Dept. of Physics, Rio de Janeiro, Brazil.
[3] Florida State Univ., Institute of Molecular Biophysics, Tallahassee, FL 32306.

ABSTRACT

Selective and extensive damage can be expected to occur when ^{125}I is placed in distinct subcellular locations. Using current ICRP tables for ^{125}I, we estimate that the total energy from ^{125}I decay is densely deposited in a CHO cell nucleus by the M-XY Auger electrons and they have a range of about 41 nm. In our model of DNA damage produced by ^{125}I, the M-XY Auger electron energy range corresponds with the diameter of a solenoid fiber in chromatin structure. Cell survival studies confirm the cytotoxicity of ^{125}I decay and especially when ^{125}I is randomly incorporated into the nuclear genome (D_0=96 decays/cell) and not in other subcellular organelles such as mitochondrial DNA. However, within the nucleus, cytotoxicity is increased three fold when only 5% of the nuclear genome is labelled with ^{125}I (D_0=30 decays/cell). Incorporation of ^{125}I into only 5% of the genome was achieved by labelling the cells in the presence of aphidicolin. The mechanism of the increased radiosensitivity is not due to a predisposed state in the cells to altered DNA damage or repair as measured by the alkaline filter elution technique nor is it due to selective irradiation of replication forks. These results indicate that ^{125}I decay can be used as a powerful molecular tool for localized damage induction in DNA in intact cells.

INTRODUCTION

Cytotoxic effects are produced when radionuclides such as the Auger electron emitter ^{125}I are incorporated into the DNA of mammalian cells via the thymidine analog, iododeoxyuridine (^{125}IdU) (Burki et al., 1973).

Unfortunately, the mechanisms of cell killing are not fully understood. However, the nucleus is generally believed to contain the target for radiation-induced cell death. Support for this notion is derived from studies where Auger emitters such as ^{125}I are specifically incorporated into cellular organelles such as the plasma membrane (Warters et al., 1977), lysosomes (Hofer et al., 1975), mitochondrial DNA (Yasui and Hofer 1986) or the general cytoplasm (Kassis 1980). The decay of ^{125}I in such non-nuclear locations is found to be non-toxic.

In this report a ^{125}I microdosimetric model of ^{125}I decay as well as possible cellular and molecular explanations for intranuclear ^{125}I decay cytotoxicity are presented. A role for chromatin organization in radiation-induced cell killing by local ^{125}I decay in the nuclear genome is suggested by these studies.

MICRODOSIMETRY

^{125}I decays more than 90% of the time by electron capture, resulting in a ^{125}Te excited atom. After the vacancy is filled, the excess energy due to the transition of an electron from an outer shell to an inner shell is transmitted to another bound atomic electron, which can be emitted from the atom in an Auger emission process. This is an alternative process to the X-ray emission for atom de-excitation. The emission of an Auger electron may lead to a cascade-like electron emission that ends the atomic de-excitation process.

A large proportion of the cytotoxic effect from ^{125}I decay is thought to be due to the localized energy deposition from the Auger electron cascade (Hofer et al., 1975). Current ICRP tables (1983) list seven different Auger electrons emitted in the process of ^{125}I decay along with their respective energies and frequencies. The frequency and the calculated range are plotted relative to energy in Figure 1. The M-XY Auger electrons have the lowest energy and thus the shortest range. However, the M-XY Auger electrons have the highest frequency of emission per decay of ^{125}I. In fact, the M-XY Auger electrons have a frequency of emission per decay ten fold higher than the average frequency per decay for the other six Auger electrons.

The relative density of energy deposition for each Auger electron emitted by a point source of ^{125}I can be visualised in the computer generated graphics diagram shown in Figure 2. In the diagram each ray represents the relative maximal range of a particular Auger electron and is indicated by straight lines.

Figure 1. The range and frequency of Auger electron emission as a function of energy are plotted. Ranges were estimated according to the expression $R = 0.0431 (E + 0.367)^{1.77} - 0.007$ (Cole, 1969).

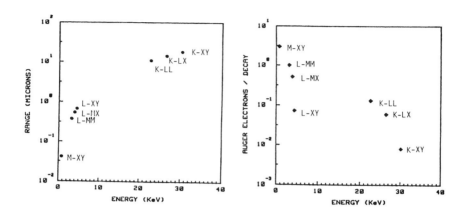

Figure 2. Computer graphics representing the density of energy deposition by Auger electrons emitted from ^{125}I decay.

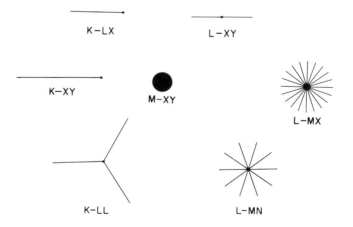

However, it should be noted that the path of each individual low energy electron undergoes an extreme degree of scatter and does not follow a straight line (Charlton, 1986; Cole,1969). Since both the frequencies and the ranges for the seven Auger electrons are spread over at least three orders of magnitude (Figure 1), each ray in Figure 2 represents the relative log of the range. The number of rays shown in Figure 2 was obtained by multiplying the frequency per decay by 20.

Experimental studies investigating the range of damage produced by ^{125}I decay (Martin and Haseltine, 1981) show that the decay of ^{125}I in synthetic linear pieces of DNA causes DNA lesions and thus produces DNA fragments. Analysis of the resulting distribution of DNA fragments indicate that the majority of the DNA lesions are produced within about five nucleotide base pairs from the site of ^{125}IdU delay. Five nucleotides would span a distance of approximately 2 nm. This difference in experimentally obtained range with calculated ranges could be explained by electron straggling and backscatter or electron emissions with ranges shorter than the M-XY auger electrons.

Figure 3. Chromatin organization relative to the energy deposition density of the M-XY Auger electrons from ^{125}I decay.

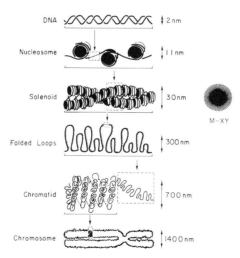

Comparison of the modelled energy deposition density with chromatin conformation indicate that the decay of DNA associated ^{125}I should cause significant irradiation of structures the size of a solenoid fiber (Thoma et al., 1979) or smaller (Figure 3). This awaits experimental confirmation. Correlation of chromatin structure dimensions with the physical decay characteristics of Auger electrons from the decay of ^{125}I suggests that extensive damage to chromatin is likely to occur. Therefore, if ^{125}I were selectively placed in such a structure, selective and extensive damage to that subcellular structure would be expected to occur. Clearly, the majority of the damaging events resulting from ^{125}I decay must occur in a highly local volume in chromatin structures.

CHROMATIN CONFORMATION

The notion that chromatin organization plays a central role in radiation-induced cell killing is further indicated by the following studies. In initial studies, the isoleucine deprivation technique (Tobey, 1973) was used to synchronize populations of Chinese hamster ovary (CHO) cells in the G1 phase of the cell cycle. The isoleucine

Figure 4. Chromatin organization of untreated control cells (A) and ile⁻ cells (B) as visualized by transmission electron microscopy.

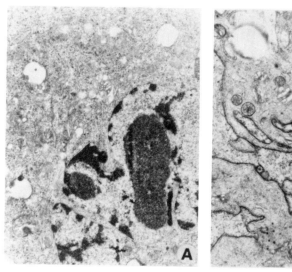

deprived cells (ile⁻ cells) were found to have decondensed chromatin (Figure 4). Further, the ile⁻ cell chromatin was observed to be more accessible to Micrococcal nuclease digestion and propidium iodide intercalation. Also, morphometric analysis of the electron micrographs indicated that heterochromatin is decondensed in the ile⁻ cells (Yasui et al., 1987). In addition, as visualized by electron microscopy, the ile⁻ cellular chromatin remains decondensed for up to eight hours after the cells are released from the G1 block, even after adding complete medium to the cells.

The isoleucine deprivation does not intrinsically change the cellular radiation response. DNA damage induction by X-irradiation, as measured by the alkaline filter elution

Figure 5. Cell survival after accumulation of ^{125}I decays at -70°C. Randomly labeled control cells (●) are compared to ile⁻ cells labelled in the absence (□) or presence of apc (▲).

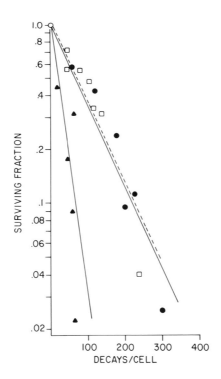

technique, does not change in ile⁻ cells over the untreated control cells. Also, the cellular clonogenic response to radiation is not different in ile⁻ cells from the response of the control cell population. Thus, the isoleucine deprivation treatment itself does not predispose the cells to enhanced sensitivity to radiation in general (Yasui et al., 1987).

The incorporation pattern of ^{125}IdU into cellular DNA was manipulated using the drug aphidicolin (apc) to direct incorporation of ^{125}IdU into a small fraction of the nuclear genome located at the nuclear matrix (Yasui et al., 1985). Treatment of cells with apc leads to the rapid decline of precursor incorporation into nuclear DNA (Yasui et al., 1985) where incorporation is reduced to only 5% of that observed for the untreated control cells.

Cells having decondensed chromatin (treated by isoleucine deprivation) subsequently incorporated ^{125}IdU in the presence of apc. These cells show increased sensitivity to ^{125}I decay (D_0 = 30 decays/cell) over cells also presynchronized using the isoleucine deprivation technique but randomly labelled in their nuclear genome with ^{125}IdU in the absence of apc (D_0 of 96 decays per cell) (Yasui et al., 1985). In comparison, cells that were not presynchronized by an isoleucine deprivation step but instead were allowed to incorporate ^{125}IdU in the presence of apc showed sensitivity to ^{125}I decay equivalent to that observed for the randomly labelled control cells (D_0 = 100 decays per cell) (Hofer and Warters, 1985).

The enhanced sensitivity of the cells to ^{125}I incorporated into the nuclear genome in the presence of apc cannot be attributed to the drug treatment itself. The apc treatment does not modify cell X-radiation sensitivity nor is this treatment cytotoxic even up to an 18 hr drug treatment (Yasui et al., 1985). In addition, the enhanced cytotoxicity cannot be attributed to the ^{125}I decays in mitochondrial DNA (Yasui and Hofer, 1986). However, the apc treatment does produce an inhomogeneous nuclear distribution of ^{125}I as measured by electron microscope autoradiography. Also the ^{125}I that is incorporated into the cellular genome is found to be locally enriched at the nuclear matrix when the structures containing supercoiled loops of DNA attached to the nuclear matrix are isolated. Thus, a three-fold increase in sensitivity to the ^{125}I decay is found for the apc treated cells which incorporated ^{125}IdU in the nuclear genome next to the nuclear matrix in an inhomogeneous distribution. The observed increase in radiation-sensitivity to ^{125}I decay suggests that not all of the nuclear genome is equally sensitive to ^{125}I decay.

The role of the particular genomic sequence labelled with ^{125}IdU in the presence of apc was investigated. Hofer and Warters, (1985) found that apc treatment permitted the incorporation of ^{125}IdU into the replication fork. However, the cytotoxicity of ^{125}I decay in the replication fork was determined to be the same as ^{125}I incorporated uniformly into the entire genome (Do = 100 decays per cell for both cases). These data rule out a sensitive genomic sequence model and indicate that the chromatin reorganization of the ile$^-$ cells subjected to isoleucine deprivation must be a critical factor in the cellular response to ^{125}I decay.

In summary, these data indicate that not all ^{125}I decays are equally radiotoxic. As previously proposed by Burki (1976), several possible mechanisms could explain a differential cytotoxicity from ^{125}I decay in the same cell line. These include 1) altered DNA damage induction or 2) DNA damage repair is modified, 3) the local proximity of the ^{125}I decay to a putative critical target is altered, 4) important genes must be irradiated to kill the cell, and 5) a molecular process associated with ^{125}IdU toxicity may be modified. The data presented in this report and the microdosimetric model together implicate chromatin compaction or chromatin organization in the cytotoxic endpoint of ^{125}I decay. Indeed, chromatin organization could alter each one of the mechanistic endpoints proposed by Burki, (1976) and thus alter cytotoxicicty. In conclusion, the data suggest that although the decay of DNA associated ^{125}I is known to produce highly toxic effects, the exact magnitude of the effect very likley depends on the location of ^{125}I in chromatin, the local conformation of chromatin, the DNA damage induction by ^{125}I decay and DNA damage repair.

ACKNOWLEDGEMENTS

This work has been supported by PHS grants NIH CA 25957, CA 45011 and CA 09097.

REFERENCES

Burki, H.J., 1976, Journal of Molecular Biology, 103, 599-610.
Burki, H.J., Roots, R., Feinendegen, L.E. and Bond, V.P., 1973, International Journal of Radiation Biology, 24, 363-375.
Charlton, D.E., 1986, Radiation Research, 109, 78-89.
Cole, A., 1969, Radiation Research, 38, 7-33.
Hofer, K.G. and Warters, R.L., 1985, Radiation Environment Biophysics, 24, 161-174.
Hofer, K.G., Keough, G. and Smith, J.M., 1977, Current

Topics in Radiation Research Quarterly, 12, 335-345.

Hofer, K.G., Harris, C.R. and Smith, J.M., 1975, International Journal of Radiation Biology, 28, 225-241.

ICRP: Radionuclide transformations - energy and intensity of emissions. Annals of the ICRP. 1983, ICRP publication 38, vols. 11-13.

Kassis, A.I., Adelstein, S.J. and Haydock, D., 1980, Radiation Research, 84, 407-425.

Martin, R.F. and Haseltine, W.A., 1981, Science, 213, 896-898.

Thoma, F., Koffer, Th. and Klug, A., 1979, Journal of Cell Biology, 83, 403-427.

Tobey, R.A., 1973, Methods in Cell Biology, 6, 67-112.

Warters, R.L., Hofer, K.G., Harris, C.R. and Smith, J.M., 1977, Current Topics in Radiation Research Quarterly, 12, 389-407. ;

Yasui, L.S., Hofer, K.G. and Warters, R.L., 1985, Radiation Research, 102, 106-118.

Yasui, L.S. and Hofer, K.G., 1986, International Journal of Radiation Biology, 49, 601-610.

Yasui, L.S., Higashikubo, R. and Warters, R.L., 1987, Radiation Research, (in press).

INDEX

The letter 'p' following the page number indicates that the term will be found on the following pages within the same paper.

Aberrations	
chromosomal	70p, 124p, 148p
Anti-cancer agents	39
Aphidocolin	187
Auger enhancement	140p
Bacteriophage	
inactivation of,	136p
T1	137p
T4	118
T7	102p
Bromine	
K-edge	113p, 136p
enhancement by,	119, 124p, 140p
enhancement ratio	119p
Bromine-77	5, 33
Carcinogenesis	160p
Cell cycle	
effects of	8, 57, 70p
G1 phase	185p
S phase	51, 71p, 82p
synchronisation	170p, 185
Cell killing	4p,18p,52p,70p,124p,136,170p, 182p

Cells
    A549   83p
    CC3 tumour   53
    Chinese hamster ovary,   8, 52, 62, 124p, 149p, 160p 185p
    CMT-93   41p
    CV1   83p
    D. Radiodurans   120
    drug resistance in   51,
    E.coli   138, 170p
    Ehrlich ascites   118
    GC   149p
    GH1   149p
    HeLa   120, 124p
    hexaploid   71p
    HEp2   42
    K562   62
    microcomputer model for, 82p
    mouse L   70p
    neoplastic   39
    oocytes   17p
    Saccharomyces cerevisiae 137p
    spermatogonial   16p, 34p
    SV40 transformed   161p
    tetraploid   71p
    V79   2, 33p, 70p, 118p, 170
    yeast   118, 137p
Chromatin   83, 148p
    conformation,   182p
    structure   59, 132
Chromium-51   1p, 33
Chromosome aberrations   70, 124p, 148p
    acentric fragments   126, 149
    isochromatid   126p
    ring and dicentric   76, 125p
Codes
    for track structure   92p
    for electron transport 30
Copper-64   83p
Copper-67   83p
Coster-Kronig electrons   2, 28, 41p

Direct effect   102, 174
Disease
    metastatic   39
DNA
    alpha-RI   60p
    break rejoining   177p

DNA (Cont.)
    conformation   108
    deletions   59p
    double-strand breaks   56p, 70p, 90p, 102p, 114p, 177p
    energy localisation in,   106
    footprinting   58
    intercalation   10, 33
    ligands   58p
    mitochondrial   187
    model for,   92
    multihit damage   131, 177
    open state of,   103
    plasmid   56p, 102p, 136p, 161
    repair of,   177p
    role of copper in,   83p
    role of zinc in,   83p
    satellite   86
    sequence selective
      binding   58
    single-strand breaks   58, 90p, 115, 137p, 171
    strand breaks   150p
    synthesis   51p
    unifilar labelling   124, 170p
Dose response
    model for   71p
Drug resistance   51
Drugs
    anti-cancer   39,
    cytotoxic   51p
    endoradiotherapeutic   41p
    Hoechst 33258   58p

Electron capture   1, 28, 40, 56p, 81p
Electrons
    Coster-Kronig   9, 28, 41p
Endoradiotherapy   39p
Energy Transfer
    in DNA   102p
Excitons   102p

Gadolinium-157   65p
Gallium-67   63p
    desferrioxamine   63
Gene amplification   160p
Genes
    Harvey-ras   160p
    hprt   62p, 171p

Genes (Cont.)
    alpha-actin   162
    Kirsten-<u>ras</u>   160p

Hoechst 33258   58
    iodo-derivative   58p
Hormone binding   148
Hybridisation   60, 161
Hydroxyurea   52p

Indirect effect   131, 174
Indium-111   1, 19p, 28p, 33
Internal conversion   1, 28
Iodine-123   1
Iodine-131   83p
Iron-55   19, 30
Iron-59   19

K-edge   112p, 124p, 136, 140p

Methotrexate   53
Micronucleii   150p
Mini-deletion
    in DNA   59p
MIRD   2, 17, 27
Mitotic delay   7
Mutagenesis   169p 180

Oncogene activation   160p
Oogenesis   16p
Ouabain   6

Phosphorus
    Auger cascades in,   120
Photon Factory   125, 137
Platinum-193m   28p
Platinum-165m   28p
Ploidy
    effect of,   71p

Radiation action
    model for   73p
RBE   10p, 19p, 33p, 39p, 91p, 169p
Repair
    of damage to DNA   177p
Resonance absorption   124p, 121

Selenium-75   1p, 19p, 33

Soliton	102p
Southern transfer	63, 161p, 179
Sperm head survival	16p,
Spermatogenesis	16p, 34p
Sulphur-35	19p
SV40	161p
Swiss Webster mice	17
Synchrotron radiation	123p, 136p
Tamoxifen	57
Target size	
dependence on ploidy	70p
Technetium-99m	1p, 18p
Thallium-201	1p, 18p, 25, 33
Thallium-204	18p, 25
Thermal neutron capture	57p
Threshold energy	98p
Thyroid hormones	148p
Thyroxine	148p
Track structure codes	30, 89p, 114p
Transformation	
by SV40	59
malignant	39p, 165
transport code	
for electrons	30
Triiodothyronine	148p
X-rays	
characteristic	90p, 112p
monochromatic	124p, 136p
Zinc 65	83p